Hubble's Legacy

Hubble's Legacy

Reflections by Those Who Dreamed It, Built It, and Observed the Universe with It

Edited by
Roger D. Launius and David H. DeVorkin

A Smithsonian Contribution to Knowledge

Smithsonian Institution
Scholarly Press

Washington DC
2015

ABSTRACT

Launius, Roger D., and David H. DeVorkin, editors. Hubble's Legacy: Reflections by Those Who Dreamed It, Built It, and Observed the Universe with It. xvi + 220 pages, 78 figures, 2014.—The development and operation of the Hubble Space Telescope (HST) have resulted in many rich legacies, most particularly in science and technology—but in culture as well. It is also the first telescope in space that has been utilized as effectively as if it were situated on a mountaintop here on earth, accessible for repair and improvement when needed. This book, which includes contributions from historians of science, key scientists and administrators, and one of the principal astronauts who led many of the servicing missions, is meant to capture the history of this iconic instrument. The book covers three basic phases of HST's history and legacy: (1) conceiving and selling the idea of a large orbiting optical telescope to astronomers, the National Aeronautics and Space Administration, and the U.S. Congress, its creation as the HST, and its definition as a serviceable mission; (2) its launch, the discovery of the flawed mirror, the engineering of the mirror fix, subsequent servicing missions, decisions on upgrades, and the controversy over a "final" servicing mission; and (3) HST's public image after launch—how the mirror fix changed its public image, how the HST then changed the way we visualize the universe, and how the public saved the final HST servicing mission. Collectively, this work offers a measured assessment of the HST and its contributions to science over more than 23 years. It brings together contributions from scholars, engineers, scientists, and astronauts to form an integrated story and to assess the long-term results from the mission.

Cover images from NASA. Front: Hubble Space Telescope orbiting Earth (see Figure 1.0-2). *Back*: Star cluster NGC 602 in the Small Magellanic Cloud, captured with Hubble's Advanced Camera for Surveys in 2004 (see Figure 3.10-3).

Published by SMITHSONIAN INSTITUTION SCHOLARLY PRESS
P.O. Box 37012, MRC 957, Washington, D.C. 20013-7012 / www.scholarlypress.si.edu

Originally published as a federal proceedings volume

Library of Congress Cataloging-in-Publication Data:
Hubble's legacy : reflections by those who dreamed it, built it, and observed the universe with it / edited by Roger D. Launius and David H. DeVorkin.
    pages cm — (A Smithsonian contribution to knowledge)
  Includes bibliographical references and index.
1. Hubble Space Telescope (Spacecraft)—History. 2. Space astronomy—History. I. Launius, Roger D., editor of compilation. II. DeVorkin, David H., 1944– editor of compilation. III. Series: Smithsonian contribution to knowledge.
  QB500.268.H825 2014
  522'.2919—dc23   2013044559
Printed in the United States of America

ISBN 978-1-935623-74-8

⊚ The paper used in this publication meets the minimum requirements of the American National Standard for Permanence of Paper for Printed Library Materials Z39.48–1992.

Contents

Preface

In the fall of 2009, the National Air and Space Museum (NASM) opened a new experimental venue in the eastern corner of its first-floor public area: a hybrid exhibit area titled "Moving beyond Earth." It provided a highly versatile and much-needed platform combining Shuttle-era exhibitry with live-stage programming, webcasting, and television production capability. Designed for a wide range of special presentations—from lectures and interactive learning experiences to live performances with themes related to the experience of air and space flight—one of its first formal efforts was to produce a special public symposium titled "Hubble's Legacy," which took place on 18 November 2009.

We timed the symposium, supported in kind by the National Aeronautics and Space Administration (NASA) in conjunction with Ball Aerospace and Technologies Corporation, to highlight the opening of two exhibits devoted to the major Hubble Space Telescope (Hubble) instruments that returned from the fifth Space Shuttle servicing mission, which had occurred the previous spring. The symposium brought together historians, astronomers, and key names in Hubble's history and legacy who were largely responsible for executing one or more of these three overlapping phases of Hubble's story:

1. The conceiving and selling of a large orbiting optical telescope to astronomers, NASA, and the U.S. Congress; its creation as the Hubble Space Telescope; and its definition as a serviceable mission.
2. Its launch, the discovery of a flawed mirror, the engineering of the mirror fix, subsequent servicing missions, decisions on upgrades, and the controversy over a "final" servicing mission.
3. Hubble's public image after launch; how the fix changed its public image; how Hubble then changed the way we visualize the universe; and how the public saved the final Hubble servicing mission.

The day began with a press conference announcing the new exhibits. Then each paper session consisted of short introductory statements by four panelists followed by a few questions panelists addressed to one another.

Each session then concluded with open questions and answers stimulated and mediated by a noted historian of space history. All proceedings were audio recorded and transcribed. The speakers were invited to elaborate on their remarks, and the editors of each section incorporated elements of the question and answer sessions into the present text.

We are particularly indebted to historians Robert Smith, Joseph Tatarewicz, and Steven Dick for their coordination and editing of each of the three sections. The editors then endeavored to give the overall manuscript a common voice.

Two of the original participants who shaped Hubble are no longer with us and so could not be included. Lyman Spitzer Jr. (1914–1997) and John Bahcall (1934–2005) were leading astrophysicists who advocated for the study of astronomy from space and who became driving forces behind Hubble's development and design. The infrared "Great Observatory," launched by NASA in 2003, the last in the series that included Hubble, was named in Spitzer's memory. Bahcall and Spitzer had coauthored a highly accessible and descriptive narrative for *Scientific American* in 1982 (see Selected Bibliography) on the state of the design, development, and promise of the Space Telescope (as it was called at the time). This benchmark article effectively laid out the rationale for, and plans relating to, the construction of an orbital space telescope; and it beautifully illustrated arguments in vogue at that time for the benefits of deploying a space telescope when some convincing was still needed. The editors highly recommend that readers consult the original article for perspective on the long journey to making the Hubble Space Telescope a reality.

Robert Smith helped to organize the first session on the conceiving and selling of the space telescope. He is a former curator and chair of the Division of Space History at NASM, where he prepared and published the prize-winning definitive history *The Space Telescope: A Study of NASA, Science, Technology, and Politics*. Robert is now a professor of History and Classics at the University of Alberta, where his principal research interest is in the history of the physical sciences with a focus on astronomy and cosmology. In his introduction to Part 1 of this volume, Robert focused on the power of an idea, introducing just how complex the process was of turning an idea such as Hubble into reality. Beyond the scientific questions, the process involved political, social, and economic factors "that posed problems just as vexing as the technical ones."

Nancy Grace Roman, a pioneer not only in space astronomy but for women in science as well, presented the first paper. After working in radio astronomy at the U.S. Naval Research Laboratory, in early 1959 Roman joined NASA, where she created the first NASA astronomical program and was named the first chief of astronomy in the Office of Space Science. As director of this program, Roman was responsible for the launch of several X-ray, ultraviolet, and gamma-ray satellite observatories. In her contribution she provides perspective on the early years of astronomy at NASA and the reaction of astronomers generally to the idea of building and launching a large optical imaging telescope.

C. Robert O'Dell, an observational astrophysicist, joined NASA as a project scientist in 1972 after being director of the Yerkes Observatory of the University of Chicago. He remained with NASA throughout the pre-Hubble phase when the preliminary designs and project pitches to the U.S. Congress were being made. During the construction of the Hubble program, O'Dell moved to Rice University and later to Vanderbilt University as distinguished research scientist in the Physics and Astronomy Department. His essay reflects on changing conceptions of operating a large orbiting optical telescope.

Edward J. Weiler has long been a voice and advocate for the Hubble program within the astronomy community at large. Recently retired as NASA's associate administrator for Space Science, Weiler had been chief scientist for the Hubble Space Telescope for almost two decades. Prior to that post he served at the Goddard Space Flight Center as the director of space operations of the Orbiting Astronomical Observatory 3 (Copernicus) and went on to become the chief of the Ultraviolet–Visible and Gravitational Astrophysics division at NASA Headquarters. He is author of the recent popular work, *Hubble: A Journey through Space and Time* (Abrams 2010), and takes particular pride in his role in planning Hubble as a serviceable national observatory. Here he recounts the campaign waged by astronauts to make it serviceable and the dramatic repair of the instrument this allowed.

Part 2 of this volume, introduced by Joseph Tatarewicz, deals with detecting and analyzing the flaw in the Hubble mirror after launch and the effort to service the instrument and return it to proper operations. Tatarewicz, now of the University of Maryland, Baltimore County, had been responsible for acquiring the Hubble Structural Dynamics Test Vehicle for NASM. Here he provides perspective on his experience presenting Hubble to the public at the museum, especially during the poignant time when the flaw was first found and the first servicing mission transformed into a rescue mission.

Three other perspectives from participants in the Hubble servicing missions follow in Part 2. The first is by John Trauger, a senior research scientist with the Jet Propulsion Laboratory and principal investigator in preparing the modifications and improvements for the successor to the Wide Field Planetary Camera (WFPC), the WFPC2. Here he recounts the painful process of determining just what the source of error was in the telescope that produced less than perfect star-like images and then determining how that error could be nullified by sending up a modified WFPC incorporating compensating optics.

Harold Reitsema is an astronomer and key member of the Ball Aerospace and Technologies Corp. staff that developed many of the scientific instruments flown on Hubble. He was called upon to build the COSTAR (Corrective Optics Space Telescope Axial Replacement) device, the instrument flown on the first servicing mission to compensate for the flawed mirror for all the instruments other than WFPC. Here he recounts the dramatic steps the Ball team took designing and building an instrument that could be flown within two years time.

John Grunsfeld is a veteran of five spaceflights. He performed several servicing and upgrade missions for Hubble and has been on missions using the Astro-2 Observatory to observe ultraviolet spectra of faint astronomical sources. He logged more than fifty-eight days in space, including fifty hours of extravehicular activity and eight space walks. As an astronomer, Grunsfeld has explored the realms of X-ray and gamma-ray astronomy and the development of detectors capable of sensing the highest known energy realms. Trained at the Massachusetts Institute of Technology and University of Chicago, he also has been a senior research fellow at Caltech (the California Institute of Technology) and observed with NASA's Compton Gamma Ray Observatory. Here, he recounts his inflight experiences and insights gained as "Chief Hubble Hugger."

During the process of preparing these essays for publication, we looked for ways to flesh out various aspects of Hubble's legacy that could not be addressed during the single day of paper sessions. Related to the essays in Part 2, and appearing here as the Appendix, is an internal technical report prepared by former NASA chief historian Steven J. Dick. It is an assessment of the decision to forego one last servicing mission for the Hubble in the aftermath of the *Columbia* accident on 1 February 2003. The NASA administrator at the time, Sean O'Keefe, was skittish when it came to risking astronaut lives on another servicing mission. After investigation, O'Keefe determined that he would allow Hubble to end its operational life earlier than expected. Dick studied this decision extensively and presents the story of how the mission was cancelled and then reinstated in 2003–2005 by a new NASA administrator.

Part 3 of this book, introduced by Steven J. Dick, explores Hubble's scientific and cultural impact. The four contributors to this part passionately and dramatically illustrate how the Hubble mission was restored and how it subsequently transformed our view of the universe. Kenneth Sembach, head of the Hubble Mission Office at the Space Telescope Science Institute (STScI), recalls the painstaking steps in the processes of bringing Hubble back into operation after the first critical repair mission. He also describes the process through which astronomers gained access to Hubble.

David Leckrone, an astrophysicist and former chief scientist for Hubble, presents a highly personal view of the extraordinary power of Hubble and its place in the history of astronomical telescopes of the past four centuries. He describes how it has captured the imaginations of scientists and the public and characterizes Hubble's scientific legacy.

Zoltan Levay, of the Johns Hopkins University Space Telescope Science Institute and a member of the Hubble Heritage Team, describes the process whereby Hubble data are transformed into images and other forms of media for public and educational consumption. He recounts how the imagery has become ubiquitous through the techniques that STScI specialists employed following aesthetic protocols used by artists of past generations. His presentation helps to illustrate just why the Hubble images are so compelling, a question further illustrated and amplified by art historian Elizabeth Kessler, formerly of Ursinus College, who focuses on the aesthetic techniques employed by Hubble artists. She argues that these techniques have their antecedents in art history and thereby shows that the products of Hubble imagery are a form of aesthetic persuasion.

The historians contributing to this volume have, during Hubble's lifetime, been chiefly interested in preserving and understanding its legacy to science, to society, and to themselves. The scientists contributing to this volume have tried to communicate their personal excitement, frustration, and devotion to what has taken up a large part of their lives and has become so much a part of their professional identities. Underneath the science they leave to posterity and the insights into what stimulates them as "Hubble Huggers" lies the material legacy of the project. Some of them have carefully and lovingly saved bits and pieces from the trash bin that are especially meaningful to them. Others have seen to it that important relics are saved and made available to those who care for the "congealed culture" of our nation. As one of those caretakers, or curators, David DeVorkin reflects on this effort, describing NASM's support for identifying, collecting, preserving, and interpreting Hubble's material legacy.

Whenever scholars take on a project such as this, they build on the efforts of earlier investigators and incur a good many intellectual debts. The editors and authors acknowledge the assistance of the institutions and individuals who aided in the preparation of this book. Our greatest debt is to Ball Aerospace and Technologies Corp. for providing the generous grant that made possible the symposium on which this volume is based. We also acknowledge the many people at the Smithsonian Institution who supported this endeavor. The symposium was planned at NASM, and we acknowledge the support of the staff in completing this effort.

For their many contributions in completing this project we also thank Jane Odom and her staff archivists at the NASA History Office who helped track down information and correct inconsistencies. In addition to Steve Dick, Steve Garber, and Nadine Andreassen at NASA, the staffs of the NASA Library and the Scientific and Technical Information Program provided assistance in locating materials. Marilyn Graskowiak and her staff at NASM Archives, and many archivists and scholars throughout several other organizations, generously lent time and expertise in locating materials. Patricia Graboske, head of publications at NASM, provided important guidance for this project, as did editorial production personnel of Smithsonian Institution Scholarly Press--Publications Specialist Meredith McQuoid-Greason and freelance editor Eva Silverfine-Ott. Our deep thanks are due to all of these fine people.

In addition to these individuals, we acknowledge the following individuals for a variety of reasons: Bruce Campbell, General John R. Dailey, Jean DeStefano, Jens Feeley, Lori B. Garver, James Garvin, John Grant, G. Michael Green, D. Wes Huntress, Peter Jakab, Violet Jones-Bruce, Sylvia K. Kraemer, John Krige, Jennifer Levasseur, John M. Logsdon, Jonathan C. McDowell, Karen McNamara, Valerie Neal, Allan A. Needell, Michael J. Neufeld, Alan Stern, Harley Thronson, and Margaret Weitekamp. Several interns provided assistance at various stages of this project, and to them we offer our sincere thanks: Lauren Binger, Jessica Bradt, Kate Carroll, Kerrie Gensch, Dina Green, Jessica Kirsch, Vicky Lindsey, Alina Naujokaitis, Amanda Peacock, Heather van Werkhooven, and Helen Yamamoto.

As is naturally the case with efforts at contemporary documentation, these essays are offered as reflections of the impressions of key participants and observers who were at or near the scene of action. We have tried to retain a sense of the passion each presenter holds for the subject, moderating their personal voices only enough to provide balance throughout the volume. We selected these voices because they are advocates largely from the world of science rather than solely from a NASA perspective. We keenly know that not all informed readers will agree with every statement or gesture. Our purpose overall has been to preserve and make accessible the personalities who played leading roles in creating Hubble's Legacy.

Introduction

Roger D. Launius and David H. DeVorkin

The Hubble Space Telescope (Hubble) has many legacies. There is a rich scientific legacy (the wealth of astronomical information it has added and continues to add to our store of knowledge about the universe); a cultural legacy (our new vision of the universe and the menagerie of fantastic things that exist within it, beyond any comprehensible scale or human reference, but still somehow brought within our grasp); and a technological legacy. Additionally it is the first telescope in space that has been repeatedly visited for repair and for improvement.

That Hubble could be visited repeatedly made it different from all other telescopes heretofore placed into orbit. It was not the first to be serviced. The Solar Maximum Mission (SMM) in the late 1980s, and certainly the solar telescopes aboard the Apollo Telescope Mount connected to the Skylab space station in the early 1970s, were visited and operated during their lifetimes in space. The SMM, in fact, was rejuvenated as well, but the extent of the transformation made possible by successive visits to Hubble was far more significant, vastly increasing and broadening its capabilities to observe the universe.

The National Aeronautics and Space Administration (NASA) began research in the latter part of the 1960s on a suitable pointing system for the large assemblage that was originally recommended in 1965 by the Space Science Board and eventually became Hubble. Initially NASA planned to build a four-meter mirror, the largest mirror that would fit inside the shroud of a Saturn upper stage. Later, to take advantage of existing national security optical production facilities, NASA reduced the size of the mirror to 3.2 meters. With the beginning of Space Shuttle flights in the 1980s, NASA redesigned the telescope with a 2.4-meter mirror so that it could be launched by the shuttle and serviced in space. Ultimately the $2 billion Hubble was launched from the Space Shuttle in April 1990, and astronomers were excited that it represented a quantum leap forward in astronomical capability. They expected to see objects in deep space with much greater optical resolution than ever before, scrutinizing the details of galaxies over 10 billion light years away. The 2.4-meter mirror was the key component: it would be responsible for collecting the light. It needed to be precision-ground and shaped to an ultrafine figure from ultra-low-expansion borosilicate glass, and then coated with a thin aluminum–magnesium fluoride reflecting surface to produce the finest detail optical laws allowed.

Pent-up expectation and anticipation followed the launch and deployment of Hubble, as astronomers waited for the first images to be processed from the data stream returning to Earth. The first images seemed bright and crisp, set against the black background of space. They were indeed clearer than pictures of the same targets taken by ground-based telescopes. Controllers then began moving the telescope's mirrors to better focus the images. Although the focus sharpened slightly, the best images still had a pinpoint of light encircled by a hazy ring, or halo. This was just not right.

Within a few weeks, NASA technicians, advised by somber astronomers, concluded that the primary mirror suffered from what optical experts call "spherical aberration." Not apparent to casual inspection, this defect amounted to a slight deviation from a perfect optical figure by only one-fiftieth the width of a human hair. But this was more than enough to prevent Hubble from focusing all light to a single point. It was not a question of producing pretty pictures, though that would come. No, it was a question of efficiency. None of the instruments aboard Hubble would be able to work to specification unless the mirror's vision was somehow corrected.

At first many believed that the spherical aberration would cripple the 43-foot-long (~13 m) telescope, and the media howled. But scientists soon found a way with computer enhancement to partially work around the aberration, and engineers planned a shuttle repair mission to correct it with a specially designed collection of tiny corrective mirrors. These mirrors were inserted into a rectangular box that would replace one of the four axial instruments aboard Hubble.

In December 1993 NASA launched the shuttle *Endeavour* on a repair mission to insert corrective equipment into the telescope and to service other instruments. During a week-long mission, *Endeavour*'s astronauts conducted a record five spacewalks and successfully completed all programmed repairs to the telescope. Astronauts Jeff Hoffman, Storey Musgrave, Kathy Thornton, and Thomas Akers performed their tasks in full public view, replacing several components of the telescope, including an altered and improved version of its all-important imaging system, the Wide Field Planetary Camera (WFPC). This instrument was swapped out for a new camera array that compensated for the primary mirror's flaw, new solar arrays were installed, and an instrument called COSTAR (Corrective Optics Space Telescope Axial Replacement), the array of tiny mirrors that corrected the flawed vision of Hubble and fed unaberrated light to the other focal plane instruments, completed the repair. The first reports from the newly repaired telescope indicated that the images being returned were more than an order of magnitude clearer than those obtained before.[1]

By early 1994 Hubble was returning impressive scientific data on a routine basis. For instance, as recently as 1980 astronomers had believed that an astronomical grouping known as R-136 was a single star, but Hubble showed that it was made up of more than 60 of the youngest and heaviest stars ever viewed. The dense cluster, located within the Large Magellanic Cloud, is about 160,000 light years from Earth, or over 900 quadrillion miles (~1,500 quadrillion km) away.

Because of the success of the first servicing mission, Hubble dominated media attention devoted to space science activities. Results from Hubble during subsequent years touched on some of the most fundamental astronomical questions of the twentieth century, including the existence of black holes and the age of the universe. In 1995, Space Times, the magazine of the American Astronautical Society, heralded the accomplishments of the first year:

- Compelling evidence for a supermassive black hole in the center of a giant elliptical galaxy located over 50 million light years away
- Observations of great pancake-shaped disks of dust, raw material for planet formation, swirling around at least half of the stars embedded in the Orion Nebula, hinting that the processes which may form planets is common in the universe
- Confirmation of a critical prediction of the Big Bang theory, that the chemical element helium should be widespread in the early universe
- Announcement by astronomers in October 1994 of measurements that limited the expansion age of the universe to be between 8 and 12 billion years old, setting the stage for precision cosmology

These measurements were the first step in a three-year systematic program to measure accurately the scale, size, and age of the universe.[2]

And discoveries continued to flow thereafter. For instance, scientists using Hubble obtained the clearest images yet of galaxies that formed when the universe was a fraction of its current age. Hubble also revealed new features on the planets, imaged the Eagle Nebula in search of information about star formation, and observed the spectacular crash of Comet Shoemaker–Levy 9 into the planet Jupiter in 1994.

Since that first servicing mission, Hubble was visited several more times to repair various functions and to replace instruments with upgraded ones that have new capabilities benefitting from rapidly improving electronic solid state detector technologies. After each mission, Hubble's vision and versatility became more acute. Hubble continually produced new data to address scientific questions, and astronomers soon found that this material could be artfully crafted into spectacular new views of the universe and all the strange and wonderful things that reside within. Largely through the efforts of astronomers and the Hubble Heritage Team, a new visible universe started to appear—on magazine covers, clothing, beer mugs, and baseball caps. Hubble became a media darling.

In modern astronomy, no telescope works alone. Hubble teamed up with a fleet of X-ray, gamma-ray, and infrared space telescopes, as well as giant ground-based radio and optical observatories, in a quest to catch and scrutinize rapidly breaking events from supernovae in distant galaxies to gamma-ray bursts. Gamma-ray bursts may represent the most powerful explosions in the universe since the Big Bang. Before 1997 astronomers were stumped: although they had observed more than 2,000 bursts, they couldn't determine whether these fireballs occurred in our galaxy or at remote distances. Hubble images showed unambiguously that the bursts actually reside in far-flung galaxies rife with star formation.

Hubble has also been used to undertake cooperative projects that have given astronomers deeper optical and infrared views of the universe than ever before. In a census of 27 nearby normal galaxies Hubble has found at least three of them harbor supermassive black holes. From this, some astronomers suggest that nearly all galaxies may harbor supermassive black holes that once powered quasars—extremely luminous objects in the centers of galaxies—but are now quiescent. After the loss of the Space Shuttle *Columbia* in 2003, Hubble looked to be in the last stages of its service life. Sean O'Keefe, the NASA administrator, cancelled a scheduled servicing mission in deference to completing the International Space Station (ISS) as quickly as possible. He also believed that a visit to Hubble was too risky for the astronauts, since it moved in a different orbit than did the ISS. If something went wrong during such a visit, there would be no safe harbor on the ISS. The shock of the loss of *Columbia* was palpable, and decisions made in its aftermath represented a draconian response to the tragedy. A new NASA administrator reversed this decision in 2005, however, and a new servicing was scheduled for 2009.

In anticipation of this mission and its success, the National Air and Space Museum planned a two-day public program that would provide both educational and informational access to Hubble, acquainting and sensitizing the museum's audience to the importance of the mission, its challenges, and the expected and realized rewards. Part of this programming was a day-long public symposium that brought together some of the pioneer figures who played significant roles in the history of defining, selling, building, and operating the Hubble Space Telescope. Amidst museum displays of Hubble, including a full scale Structural Dynamic Test Vehicle, the original backup primary mirror that ironically had the proper optical figure, critical elements of the original WFPC that was returned from orbit in 1993, the Faint Object Spectrograph that confirmed the existence of supermassive black holes, and the wealth of images and information about the universe that these and other components amassed, we brought together some 15 pioneer builders and users, along with noted historians, to contemplate the many legacies of Hubble. This book is, therefore, an anthology of what have been personal journeys of major participants in one of the most significant scientific quests of the twentieth and twenty-first centuries.

Throughout the process of organizing the two-day public program at the National Air and Space Museum, we kept in mind that when the Hubble Space Telescope was finally launched in 1990 after more than 20 years of design and construction, it was arguably the most complex and sophisticated spacecraft ever created. From its vantage point 300 miles (~500 km) above Earth, the 12-ton (~11-metric-ton) orbiting observatory promised to revolutionize astronomy and cosmology. But what few had anticipated, inside NASA and even in the scientific community, was its amazing ability to spark the imagination through the visions its data made possible. Here we try to capture some of that passion in the words of those responsible for making it possible and making it happen.

Notes

1   Joseph N. Tatarewicz, "The Hubble Space Telescope Servicing Mission," in From *Engineering Science to Big Science: The NACA and NASA Collier Trophy Research Project Winners*, ed. Pamela E. Mack, pp. 365–396 (Washington, D.C.: NASA SP-4219, 1998).

2   "Hubble Space Telescope Scientific Results in 1994," *Space Times: Magazine of the American Astronautical Society* 34 (March–April 1995), 11.

Part 1

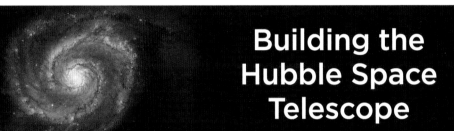

Building the Hubble Space Telescope

Introduction: The Power of an Idea

Robert W. Smith

"Cape Canaveral, Fla., April 24 [1990]—In a thunderous overture to a promised new era in astronomy," the *New York Times* reported, "the space shuttle *Discovery* rocketed into orbit today with the...Hubble Space Telescope, which scientists believe will give them a commanding view of the universe as it was, is and will be."[1] When it was launched the Hubble Space Telescope (HST) was widely reckoned to be the most powerful astronomical telescope ever sent into space. Although by the standards of ground-based astronomy in 1990 its 2.4-meter-diameter primary mirror was not especially large, the HST's orbit roughly some 600 kilometers above the Earth's surface would, it was expected, more than compensate for its size as the HST would be flying above all but a tiny fraction of the Earth's atmosphere. The HST would be going to a place where the stars do not twinkle, and it carried aloft the hopes not just of astronomers but numerous individuals who had been involved in its construction across the USA and Europe. As well, the huge wave of pre-launch publicity had heightened the interest of many members of the public who might have been hazy about HST's scientific goals but had been told it would do great things in exploring the universe and in revealing astronomical objects with unprecedented clarity.

Hubble's journey from conception to orbiting in space, however, was a complex, costly, and very drawn-out process. It involved many thousands of people (the great majority of them non-astronomers) and hundreds of different organizations that would in time expend an enormous amount of work in terms of coordination and systems engineering. Hubble's journey to space also involved far more than the framing of scientific questions and the development of technologies appropriate to an orbiting observatory. Rather, it also involved political, social, and economic factors that posed problems just as vexing as the technical ones. It saw the engagement of an assortment of institutions: the U.S. Congress and the White House; the National Aeronautics and Space Administration (NASA) Headquarters in Washington, D.C.;

NASA centers across the USA (most importantly the Marshall Space flight Center in Huntsville, Alabama); the European Space Agency, universities in North America and Europe, and, crucially, a wide range of companies that would bring the indispensable engineering and manufacturing skills that would make the design, construction, and operation of the HST feasible. Established in the early 1980s, the Space Telescope Science Institute in Baltimore would become the scientific "nerve center" for the project and would play an absolutely essential role in preparing to operate the HST in orbit. By the time the HST was launched the institute's staff had also made vital contributions in many other areas of the project too.

Thinking about space telescopes in orbit, however, predated even the creation of NASA in 1958. Some of the early space visionaries considered the possibilities of telescopes in space. Hermann Oberth, for example, in his 1923 *Die Rakete zu den Planetenräumen* (By Rocket into Planetary Space) discussed both an orbital space station in geosynchronous orbit and a telescope that could be attached to this station. Oberth's vision even predated the first successful launch of a liquid-fuelled rocket by the American rocket pioneer Robert Goddard in 1926. Goddard's small rocket rose some 41 feet (-12 m) above Auburn, Massachusetts, in a flight that lasted just two seconds. Designing a telescope to be launched into space clearly posed all sorts of daunting technical challenges in and of themselves, but without a suitable rocket even the most brilliantly conceived telescope design would be grounded.

The development of the German V-2 rocket during World War II was therefore a hugely important step on the road to launching telescopes into space. From the 41 feet of Goddard's initial liquid-fuelled rocket flight, the first successful flight of a V-2 makes a striking contrast. One afternoon in October 1942 at Peenemünde on Germany's Baltic coast, a V-2 climbed to a height of some 56 miles before diving back into the Earth's atmosphere and smashing into the Baltic over a hundred miles from the launch site. A human-built object had, for the first time, climbed to the very edge of space. The V-2 weighed around 31,000 pounds, was some 46 feet tall, was propelled skywards by the explosive mix of liquid oxygen and alcohol, and all in all was a very complicated machine. Soon these rockets were being directed at targets in England and the European mainland.[2]

For a young Yale professor of astronomy and fan of science fiction, Lyman Spitzer Jr., the startling progress of rocketry during the war changed everything. He dreamt of peaceful uses of rockets as the means to launch space telescopes. What had previously been consigned to the realm of science fiction, might, after all, be turned into sober reality.

In 1946 Spitzer (Figure 1.0-1) was a consultant for the newly established

Figure 1.0-1. Lyman Spitzer Jr. was a professor of astrophysics at Princeton University, director of the Princeton Observatory, and founder of the Princeton Plasma Physics Laboratory. The infrared "Great Observatory" launched by NASA in 2003, as the last in the series that included the HST, was named in his honor. (NASA photo.)

Project RAND (Research And Development) that had been founded by the Douglas Aircraft Corporation (in time this would evolve into the very well-known RAND Corporation, a think tank that undertook studies on behalf of the U.S. Air Force). He first discussed the atmosphere above 300 kilometers for RAND but later pondered orbiting telescopes for a study of an Earth-circling spaceship, in particular the scientific potential of such radical new technologies. Spitzer's report was entitled "Astronomical Advantages of an Extra-terrestrial Observatory." In it he explored the astronomical observations that could be made from a satellite in three different cases: (1) without a telescope, (2) with a telescope with a 10-inch-diameter mirror, and (3) with a telescope with a primary mirror between 200 and 600 inches in diameter. In 1946, the biggest telescope in operation at a ground-based observatory was the 100-inch (2.5 m) telescope perched atop the Mount Wilson Observatory in California; the great 200-inch (-5 m) reflector on Palomar Mountain was still two years away from completion. A telescope in space that boasted a 200-inch or bigger primary mirror was therefore clearly far in the future. But Spitzer judged the potential of space-based telescopes to be enormous and to promise great gains over ground-based telescopes.[3]

For astronomers, there were perhaps three major advantages to a space telescope over an equivalent ground-based instrument:

1. Improved wavelength coverage
2. Extended observing day (for a high orbit)
3. Improved angular resolution

Let's take a short detour from the early history of the HST to consider each of these advantages in turn. First, the Earth's atmosphere blocks most of the wavelengths of electromagnetic radiation that reach our planet from astronomical sources. This is certainly fortunate for us because otherwise the surface of the Earth would be bathed in harmful radiations. It is, however, a very serious handicap to astronomical investigations. Astronomers who want, for example, to explore the X-rays emitted by astronomical objects have no choice but to send their X-ray detectors into space because the Earth's atmosphere stops all X-rays. Similarly, astronomers who want to observe in the ultraviolet and large portions of the infrared wavebands need to fly their telescopes above the atmosphere. The HST (Figure 1.0-2) was designed to enable observations in the wavelength range from 120 nanometers to 1 millimeter, that is, in the optical, ultraviolet, infrared, and submillimeter wavebands. The first complement of HST scientific instruments was selected in 1978 and comprised instruments sensitive to ultraviolet and optical wavelengths. Scientific instruments inserted into the telescope later, however, equipped with vastly improved detectors, pushed the telescope's observing capabilities into the near infrared.

The second major advantage of a space telescope in a high orbit versus a big ground-based telescope was anticipated to be the 24-hour observing "day." Ground-based astronomers were used to securing their very best observations on the few nights of the year when observing conditions are at their finest. A space telescope in high orbit, however, would have superb 'seeing' conditions all the time as it would not have to contend with vagaries of the Earth's atmosphere. The quality of such a telescope's observations was therefore expected to be consistently very high.

Third, and most important, was the improved angular resolution of a space-based telescope compared with one on the ground. Angular resolution refers to the ability of a telescope to distinguish details in an astronomical image. Other things being equal, the bigger the telescope mirror, the better the angular resolution and the more details that can be discerned. Beyond

Figure 1.0-2. Hubble Space Telescope orbiting Earth after deployment on second servicing mission. (NASA image STS082-746-059.)

the obvious limitations caused by atmospheric turbulence, which makes the stars twinkle, the night sky is far from dark; in fact it is still somewhat bright even on the darkest of nights. This is mostly due to what is known as airglow. Airglow is caused by various processes high in the Earth's atmosphere. There is also an additional glow from the so-called zodiacal light caused by the scattering of sunlight from dust grains spread throughout interplanetary space. Even in space there is still a very slight glow due to the zodiacal light as well as light from stars that has been scattered by dust in interstellar space. However, in space the glow is much reduced from that experienced by a ground-based telescope. The sky background in each "resolution element" for a space telescope and its associated light detectors, therefore, would be much less than that for an equivalent ground-based telescope, significantly improving the space telescope's performance. As one astronomer wrote in 1979 (by which time the HST's primary mirror size had been set to 2.4 meters):

> The implication...is that point objects [like stars] 50 times fainter than the faintest observable from the ground can be detected by the [Hubble] Space Telescope. Roughly speaking, these objects can be observed to distances a factor of 7 times greater than at present, or, if the objects are uniformly distributed in space, 350 times more of this class of object become[s] accessible to observation.[4]

For astronomers, then, the advantages of a large telescope in space were very apparent. This does not mean, however, that astronomers rushed to be involved with space astronomy. After the end of the war, the U.S. Army launched a number of V-2 rockets that had been assembled from captured parts to gain experience with this new technology. Scientific payloads were sent aloft on many of them. Later American-built rockets began to be flown with scientific payloads. But initially, few astronomers were interested in devoting their efforts to flying instruments into space.[5] Even when things went well (and they often did not!), instruments aboard such rockets were able to secure only very limited amounts of scientific data. Such rockets carried the instruments above the atmosphere for only short amounts of time, typically around five minutes, before the rocket would start to arch back toward the Earth and plunge into the atmosphere. A satellite that orbited around the Earth, however, would not be limited in the same manner as a rocket, and so to space astronomy enthusiasts an observatory that could operate for months or years in orbit (Figure 1.0-3), or perhaps on the surface of the Moon, was the great prize. But even in the early 1950s, such a space observatory looked to be decades away.

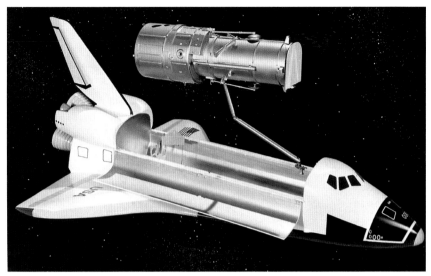

Figure 1.0-3. Artist's concept from 1980 depicts the HST being positioned for release from the Space Shuttle orbiter by the Remote Manipulator System. None of this equipment had been built at the time this rendering was done, but all were in production. (NASA illustration MSFC-75-SA-4105-2C, http://mix.msfc.nasa.gov/abstracts.php?p=1693.)

The launch of *Sputnik I* in October 1957, however, transformed hopes for launching a true space observatory into the very near future. The orbiting of the roughly basketball-sized Soviet satellite set in train events that rapidly led to what became known as the Space Race between the USA and the Soviet Union. The main theatre of the Space Race was provided by the human spaceflight program, but the newly established U.S. space agency, NASA, also embarked quickly on a vigorous program of scientific research using space vehicles. The superpower rivalry in space meant that money was soon flowing to space astronomy. By the standards of astronomy in the late 1950s and early 1960s, this was spending on a staggering scale. When compared with the standards of the immediate post World War II era it was prodigious.

Ground-based astronomy in the USA had long proven to be a remarkably attractive field for private donors. The influx of private support in the late-nineteenth and early twentieth centuries was key in the transformation of the USA from an astronomical backwater into the leading power in observational astronomy in the world in the twentieth century. It all started with the two largest refractors by 1900, a 100-inch (2.5 m) reflector that went into service on Mount Wilson in California in 1919, and a 200-inch (~5 m) telescope on Palomar Mountain that saw first light in 1948. More recently, a growing host of 8-meter reflectors and subsequently the Keck Observatory's two 10-meter telescopes on Mauna Kea in Hawaii have kept the USA at the forefront of the world's optical telescopes. The majority of these telescopes are examples of the generosity of wealthy private donors in funding state-of-the-art instruments.

Space astronomy, however, requires a different order of investment as well as organizational structure. The dominant trend after World War II in the USA was to create national facilities for science, in particular astronomy and physics. And the heady political reaction to the challenge of *Sputnik* made science a national priority, centering large space initiatives squarely within a national framework. Thus space astronomy has been entirely government funded (military and civilian) in both the USA and Europe. So the advocates in the USA of a large telescope in space recognized by the early 1960s that if their dreams were to become reality then they would need the backing of the White House and the Congress.

Economics and politics therefore loomed large. In comparison with the dollars devoted to ground-based astronomy, space astronomy was fuelled in the late 1960s by astonishingly high amounts of federal government dollars. But the different wavelength ranges were treated rather differently. In the 1960s X-ray astronomers received less funding than did ultraviolet astronomers, who struggled to meet NASA's decision to create a number technically very demanding Orbiting Astronomical Observatories (OAOs; Figure 1.0-4).

There was, then, competition for resources among different groups of scientists interested in exploring the universe via different wavelength regions. Advocates of a Large Space Telescope who were also OAO scientists, like Lyman Spitzer, had therefore to build a strong case that their favored projects should be funded. Deciding which observing wavebands and which groups of researchers to fund to study those wavebands were not straightforward choices for NASA managers.

In the 1960s, NASA battled the Soviet Union for prestige through space projects. In terms of prestige, the most effort by far, of course, went into the Apollo program, but science projects also made contributions. The then head of NASA's Space Science program, Homer E. Newell, claimed in 1966 that "for the initial exploration in virgin fields relatively simple instrumentation and limited-scope research programs often suffice to permit rapid exploration of new technologies and breakthroughs of understanding."[6] But as Newell would later recall, "exposed directly to the outside pressures to match or surpass the Soviet achievement in space, NASA moved more rapidly with the development of observatory-class satellites and the larger deep-space probes than the scientists would have required...some of the most intense conflicts between NASA and the scientific community arose later over the issue of the small and less costly projects versus the large and expensive ones—conflict

Figure 1.0-4. Assembly of the Orbiting Astronomical Observatory (OAO) for shroud jettison tests in 1965 at the Space Power Chamber (SPC) at Lewis Research Center, Cleveland, Ohio. Lewis is now known as the John H. Glenn Research Center at Lewis Field. The SPC consisted of two vacuum tanks that were created in 1962 inside the former Altitude Wind Tunnel. The OAO satellites, launched on Atlas-Centaur and Atlas-Agena rockets, were the first to allow astronomers to view the universe from above the Earth's obscuring and distorting atmosphere. (NASA image C1965-1458, http://grin.hq.nasa.gov/ABSTRACTS/GPN-2000-001446.html.)

that NASA's vigorous development of manned spaceflight exacerbated."[7] Big, in matters of prestige, could be better.

Given the major push that the OAOs had given to ultraviolet space astronomy, it was a relatively simple step for NASA to go beyond them with plans for a much larger scale ultraviolet–optical observatory. This does not mean, however, that as built the HST was more or less an extension of OAO technology. In fact, the HST's technical heritage came much more from the U.S. program of secret photoreconnaissance (or more colloquially, "spy") satellites than earlier space astronomy satellites. As George Keyworth, President Reagan's science advisor, was to put it in 1985, the HST "is new, but it draws upon technologies used in military systems."[8]

The OAO program of the 1960s and 1970s was nevertheless very important because it helped to establish a group of astronomers with interests in, and practical skills related to, ultraviolet space astronomy. But the scale of the planned space telescope meant that it would need the enthusiastic support of not just a small group of astronomers but a considerable body of the entire astronomical community in the USA. A crucial concern for advocates of what became the HST was building up their base of support among astronomers as well as within NASA.

But some ground-based astronomers worried that space astronomers were their rivals for restricted resources. That is, money might be siphoned off from ground-based astronomy to fuel space astronomy. Given the much higher costs associated with space astronomy enterprises, this was not a comforting thought. The largest and most expensive of the early space astronomy projects was, again, the OAOs, and so this project drew criticism. Many very big ground-based telescopes, ran one line of argument, could be purchased for the $200 million plus price of the three planned OAOs. The third of this series was the Copernicus satellite. It was launched in 1972 and operated for eight years. It was this satellite that served as Ed Weiler's introduction to space astronomy when he joined the staff at Princeton in 1976 and worked at the Goddard Space Flight Center on Copernicus' science operations. Copernicus carried an ultraviolet telescope (with an aperture for the primary mirror of 32 inches [-80 cm]) from Princeton University and X-ray detectors from University College, London. Now, $200 million plus in early 1960s dollars equates to about $1.5 billion in 2011 dollars. So the OAOs were, by some margin, the most costly astronomical project undertaken before the 1990s.

One of the political lessons of the 1960s for astronomers was, however, that the money spent on space astronomy could not be readily switched to ground-based astronomy. Ground-based astronomy and space astronomy, then, were not really in competition with each other. As an aside, it is worth noting here that the money spent on space astronomy in the 1960s would pale in comparison with the spending of later decades. The detailed design and construction of the HST began in 1978, and it was launched into space in 1990. As of 2011 it was still operating and has consumed around $18 billion. The HST's planned successor, the James Webb Space Telescope, is currently priced (at time of this writing) at around $10 billion, with several years to go before it's launched.

In the late 1960s, advocates (both within and outside of NASA) for a large telescope in space worked to increase interest in such an endeavor among astronomers as well as industrial contractors who would eventually be needed for detailed designs and construction. Letting contracts for various feasibility studies, therefore, enabled technical questions to be addressed while also serving to inform astronomers and companies about the project. The authors of the three essays in the first section of this book—Nancy Roman, C. R. (Bob) O'Dell, and Edward J. (Ed) Weiler—all played very significant roles in the origins of the HST.

Nancy Roman was a scientist at NASA Headquarters from 1958 to 1980. In the 1960s she judged that space telescope advocates were inclined to underestimate just how massive were the technical challenges that lay ahead. She was, however, happy to support feasibility studies "because after all, we were never going to get there if we didn't start somewhere, and this was as good a place as any to start. I just wasn't at the point where I thought we were going to be doing it in a few years."[9]

As NASA, its industrial contractors, and astronomers completed the initial feasibility studies on what was in the late 1960s and early 1970s called the Large Space Telescope (renamed the Hubble Space Telescope in 1982), the advocates began to get a somewhat better sense of what sorts of costs might be associated with such a telescope. These costs, together with debates over the place such a telescope should hold as a national priority (some in the Congress preferred that money go to satellites they expected would have more practical benefits), led to considerable opposition to the Large Space Telescope in Congress. Indeed, the name was changed in 1975 to the Space Telescope because the removal of large made the project sound less lavish.

In a very energetic lobbying campaign between 1974 and 1978, John Bahcall (a member of the Institute for Advanced Studies at Princeton) and Lyman Spitzer Jr. (Princeton University), who of course had written the "Astronomical Advantages of an Extra-Terrestrial Observatory" in 1946, assisted by many astronomer colleagues, perhaps most notably George Field at Harvard, rallied astronomers to get behind the Space Telescope. To those following the deliberations closely, on more than one occasion it looked as if the Space Telescope's opponents might cancel the project before it had seriously started. But greatly aided by the lobbying skills of the telescope's industrial contractors, strong support from the White House as well as some members of the House and Senate, and the energetic advocacy of many astronomers, the Space Telescope did survive. In 1977 the Space Telescope was formally approved as a new NASA project with the European Space Agency (ESA) on board as a minor partner.[10]

By 1977 there were, however, a number of differences from the early 1970s in the way the telescope and the program to build it had been conceived. Among these was ESA's involvement, which meant the addition of a minor partner to NASA. At the same time, the number of dedicated scientific instruments had been reduced to five (with the telescope's Fine Guidance Sensors also planned to act in effect as a sixth instrument) and the size of the telescope's primary mirror had been reduced from 3 meters to 2.4 meters. Cost cutting had been important considerations for NASA

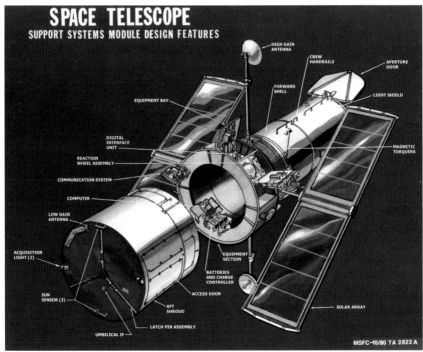

Figure 1.0-5. This illustration depicts the design features of the Space Telescope in 1980. The spacecraft is one of the three major elements of the HST (excluding the solar arrays) and encloses the other two elements—the Optical Telescope Assembly and the scientific instruments. (NASA illustration 010255, http://mix.msfc.nasa.gov/abstracts.php?p=1645.)

in all of these changes. Bob O'Dell was the NASA project scientist based at the Marshall Space Flight Center from 1972 to 1982, and so he was intimately involved in the decision-making process leading to these changes and the scientific issues related to the telescope's design. The HST would be big, roughly some 40 feet by 15 feet (-12 × 4 1/2 m), but it would need to be a precision instrument built to what was reckoned at the time to be very exacting requirements.

By the start of the detailed design and construction of the Space Telescope in 1978, the system had evolved to the stage where it had four main elements (Figure 1.0-5).

1. The Optical Telescope Assembly. This comprised the telescope itself, at the heart of which was a 2.4-meter primary mirror.
2. The scientific instruments. Designed in modular fashion so they could be exchanged in orbit by Space Shuttle astronauts, there were five dedicated scientific instruments—two cameras, two spectrographs, and a photometer—as well as Fine Guidance Sensors that could also act as a sixth instrument.
3. The solar arrays. Supplied by the ESA, the two 40 feet by 8 feet (12.2 × 2.4 m) solar arrays were designed to gather sunlight for conversion into electrical power for the operations of the spacecraft.
4. The Support Systems Module. This consisted of the spacecraft housekeeping functions minus the Optical Telescope Assembly, the scientific instruments, and the solar arrays. Its job was to provide the support—in terms of communications, electrical power, data storage, attitude sensing and control, and so on—needed for a functioning observatory.

One policy that had not been changed by the political battles of the mid-1970s was that the Space Telescope would be both launched by the Space Shuttle and regularly visited by shuttle astronauts (Figure 1.0-3). The goal was for the telescope to be both maintained in orbit and also for its scientific capabilities to be continually upgraded. As the program to build Space Telescope got underway in 1978, the aim was to call on the telescope with the shuttle about every two and one-half years. The scheme was for the telescope's set of scientific instruments to be modular in design so that astronauts could insert on orbit new instruments in place of old ones. There were also plans at this time for the Space Shuttle to fly into space, capture the telescope in its payload bay, and return it to Earth for a complete refurbishment before it was redeployed into space from the Shuttle once more. Such complete refurbishments were anticipated to take place every five years or so. The stated design lifetime for the telescope was 15 years. This, however, was not regarded as a very firm figure, and astronomers hoped that the telescope would operate for longer than that.

In time, due largely to the rising costs of Space Shuttle launches and limitations on its number of flights per year, these plans for maintenance and refurbishment were throttled back. Amid technical concerns, for example, over possible contamination of the telescope while it was on the ground, and a broader worry that if the telescope were brought to the ground it might never get back to space, plans to return the telescope to Earth periodically before re-launching it were also dropped. But as will be seen later in this volume, and as Ed Weiler discusses in his essay in this section, maintenance and refurbishment of the telescope was crucial for the telescope's long-term success on orbit. It was also essential to the HST's very survival following the discovery soon after it was launched in 1990 that a serious mistake had been made in the shaping of the telescope's primary mirror. The mirror was flawed by an optical defect known as spherical aberration.[11]

To begin, however, Nancy Roman discusses the early history in NASA of what became the HST. This is followed by C. R. (Bob) O'Dell's comments concerning the extensive effort he undertook to ensure the approval of the Space Telescope. Edward J. Weiler, the associate administrator for NASA's Science Mission directorate between 2008 and 2011, was involved with the telescope's construction as program scientist for the HST from 1979 to 1998, and he offers an assessment of the building of the HST during the 1980s.

Finally, we call attention to a seminal paper by John N. Bahcall and Lyman Spitzer Jr. that appeared in *Scientific American*[12] that reminds us of a time when the Space Telescope was envisioned but not yet a reality. Bahcall and Spitzer's benchmark article helped to set the stage for public acceptance of the HST. We heartily recommend it.

Notes

1   John Noble Wilford, "Shuttle Soars 381 Miles High, With Telescope and a Dream," *New York Times*, 25 April 1990: A1.

2   Michael J. Neufeld, *The Rocket and the Reich: Peenemünde and the Coming of the Ballistic Missile Era* (New York: The Free Press, 1995).

3   See Robert W. Smith (with contributions by Paul A. Hanle, Robert H. Kargon, and Joseph N. Tatarewicz), *The Space Telescope: A Study of NASA, Science, Technology, and Politics,* (New York: Cambridge University Press, 1989, revised edition. 1994).

4   M. S. Longair, "The Space Telescope and its Opportunities," *Royal Astronomical Society Quarterly Journal,* 20 (March 1979): 7.

5   David H. DeVorkin, *Science with a Vengeance: How the Military Created the US Space Sciences After World War II* (New York: Springer-Verlag, 1992).

6   Homer E. Newell, "Space Astronomy Program of the National Aeronautics and Space Administration," in *Astronomy in Space* (Washington, D.C.: NASA SP-127, 1967), 6.

7   Homer E. Newell, *Beyond the Atmosphere: Early Years of Space Science* (Washington, D.C.: NASA SP-4211, 1980), 97.

8   Quoted in J. Kelly Beatty, "HST and the Military Edge," *Sky and Telescope*, April 1985, 302.

9   Nancy Grace Roman, personal communication, 22 November 2009.

10   "Memorandum of Understanding between the European Space Agency and the United States National Aeronautics and Space Administration," 7 October 1977, NASA Historical Reference Collection, History Office, NASA Headquarters, Washington, D.C.

11   This aspect of the history of the HST will be discussed in Part 2 of this volume.

12   John N. Bahcall and Lyman Spitzer Jr., "The Space Telescope," *Scientific American* 247(1) (July 1982):40–51.

1

Conceiving of the Hubble Space Telescope: Personal Reflections

Nancy Grace Roman

Astronomers have always wanted to get above the atmosphere. Looking at stars through the atmosphere is not too different from looking at street-lights through a pane of old stained glass: you're limited in the colors that you can see. We are not normally aware of this limitation because our eyes are adapted to the colors that come through the atmosphere, but there are many colors that don't come through the atmosphere that tell us a great deal about the universe.

Moreover, the stained glass has flaws in it, so that the image is not sharp. The atmosphere also has flaws. It has irregularities that keep the image from being sharp and, even worse, these flaws move so that the image moves around. It's like taking a time exposure holding your camera in your hand. You can't get a sharp exposure.

Finally, the stained glass has dust on it, and the dust scatters the light so that you never get a completely dark background. Dust and, even more im-portantly, molecules in the Earth's atmosphere scatter light in the same way. That's why you don't see stars in the daytime, but scattering also happens at night. Not only is light scattered, but the atmosphere itself radiates light so that the background is never dark from the ground.

After astronaut John Grunsfeld returned from the last servicing mission in 2009, he remarked that the 2.4-meter Hubble Space Telescope (HST) mirror with the new instruments in place will to be able to reach as far as a 30-meter telescope from the ground. If that's true, it's primarily because the background is dark. Hence, it's obvious why astronomers wanted to get above the atmosphere; certainly throughout the twentieth century and possibly even before they dreamed of this possibility. In 1946 Lyman Spitzer described the advantages of a 400-inch (-10-m) telescope in space.[1]

The Space Age began with rocket programs during the Internation-al Geophysical Year (IGY) 1957–1958.[2] The National Academy of Sciences (NAS) played a major role in the planning for the IGY, so it's natural that when NASA was formed and the space possibilities became broader, the NAS played a major role in recommending what should be done in science. The NAS held a two-week summer study in 1962, in which it assembled a large number of scientists who were divided into panels according to their specialties. There were probably 10 people on the astronomy panel.[3] The discussion soon turned to the possibility of a space-based observatory. This was an exciting possibility, but some were overly optimistic about what might be accomplished.

Figure 1.1-1. Dr. Nancy Grace Roman was one of the nation's most critical astronomers in the space program throughout the 1960s and 1970s. She was central to the development of the Orbiting Solar Observatory series (illustration from 1963) as well as the Orbiting Astronomical Observatory series, and then she helped to create satellites such as the Cosmic Background Explorer and the HST. After retirement in 1979, she continued as a contractor at the Goddard Space Flight Center. Dr. Roman remains a strong advocate of women in the sciences. (NASA image 63-OSO-1, http://grin.hq.nasa.gov/AB-STRACTS/GPN-2002-000212.html.)

One participant in the 1962 meeting, an astronomer from Arizona, had looked at the Apollo rocket and had decided that the Apollo upper stage could carry a three-meter telescope. This prospect was exciting to the group, and they decided then and there that this three-meter telescope was what they wanted. I thought it was premature to pursue such a project because I was then involved with (and very much aware of the difficulties of) producing a useful Orbiting Astronomical Observatory (OAO; Figure 1.1-1). These carried much smaller mirrors and were far simpler in concept. I felt it was too early to start thinking about a satellite carrying a three-meter mirror. In fact, the OAO that was launched in 1968, which was the first successful one, carried only a battery of small telescopes, the largest with a 16-inch (0.40 m) mirror and a cluster of four 12-inch (0.30 m) mirrors. In 1972, another OAO was launched with a 32-inch (0.82 m) mirror and enjoyed a highly successful career studying the spectroscopic nature of the ultraviolet universe, but it was still a far cry from a 3-meter mirror system capable of high-precision imaging.

Opinions were mixed at that 1962 meeting. As Aden B. Meinel, director of the Optical Sciences Center at the University of Arizona, wrote about one participant's position: "Ira Bowen [the director of the Mount Wilson and Palomar Observatories] said at one meeting that one could never stabilize a space telescope enough to yield high resolution. He said that simply pulling out the dark slide would disturb it. He also remarked that higher [angular] resolution wouldn't be of much importance to astrophysics."[4]

The astronomical community therefore was initially divided over the possibilities of a large orbital space telescope. But by the 1965 Space Studies Board summer study, momentum behind the project had grown. The Langley Research Center had been responsible for NASA's human space program up to that point. Some of the engineers there jumped on the idea of developing a large, manned orbiting telescope, and they teamed up with aerospace companies and presented designs for it. This was the last thing astronomers wanted! Aside from the fact that for almost a century, with one small exception, research had not been done by a person looking through a telescope, astronomers knew a person would need an atmosphere, and that was what we were trying to get away from. In addition, a person would wiggle during long exposures, which would cause the telescope floating in orbit to wiggle in the opposite direction, blurring the image.

The aerospace companies had done some rather detailed studies of what a space telescope might consist of and what the enabling technologies for it should be, but they did so within the biases of their craft. Because the aerospace companies were almost completely unaware of what astronomers really needed, these studies were not very useful for astronomy, just as Ira Bowen long feared.[5]

For many astronomers, there was no question about the three-meter telescope in space, but at the 1965 meeting there were questions about how to design the mission. Would the telescope operate in Earth orbit or should it be on the Moon? That question remained undecided for quite a while. Additional studies helped to show that a lunar base was not as effective as an Earth orbit. The Moon might well provide a stable base, making the telescope less sensitive to the motion of parts, and also would provide a reference system for the pointing controls. Connected to a manned base, a Moon-based telescope could be used much as ground-based telescopes are used. But there were serious disadvantages with the Moon. Perhaps the most serious one was that it was unclear how soon such an installation would be feasible. The Moon appeared to be undesirably dusty. Moreover, its motion is complex, making the guidance difficult before modern computers were well developed. Nevertheless, the issue remained alive until the early 1970s.

While the site for the telescope remained debatable, Lyman Spitzer led a subcommittee of the Space Studies Board to define the science to be undertaken with a large space telescope. Through his efforts, and those of a small circle of colleagues, by 1970 the possibility of such a large orbital telescope had gained sufficient support that NASA established two committees: a Large Space Telescope (LST) task group to map out the engineering

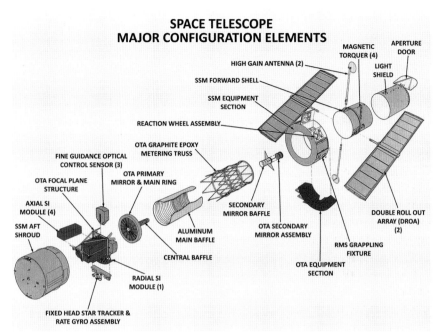

**SPACE TELESCOPE
MAJOR CONFIGURATION ELEMENTS**

MAGNETIC TORQUER (4)

APERTURE DOOR

HIGH GAIN ANTENNA (2)

LIGHT SHIELD

SSM FORWARD SHELL

SSM EQUIPMENT SECTION

REACTION WHEEL ASSEMBLY

OTA GRAPHITE EPOXY METERING TRUSS

FINE GUIDANCE OPTICAL CONTROL SENSOR (3)

OTA PRIMARY MIRROR & MAIN RING

OTA FOCAL PLANE STRUCTURE

AXIAL SI MODULE (4)

SSM AFT SHROUD

ALUMINUM MAIN BAFFLE

CENTRAL BAFFLE

RADIAL SI MODULE (1)

SECONDARY MIRROR BAFFLE

OTA SECONDARY MIRROR ASSEMBLY

DOUBLE ROLL OUT ARRAY (DROA) (2)

RMS GRAPPLING FIXTURE

OTA EQUIPMENT SECTION

FIXED HEAD STAR TRACKER & RATE GYRO ASSEMBLY

Figure 1.1-2. The Space Telescope's configuration as of January 1980. This exploded view situates the Support Systems Module (SSM), which forms an outer shell that protects all systems and provides power, communication, and control. The Optical Telescope Assembly (OTA) includes the primary and secondary mirrors, a graphite epoxy truss, and the focal plane assembly where the scientific instruments (SI) are located. The explicit servicing elements are visible, including the Remote Manipulator System (RMS) grappling fixture and the modularity of both the axial and radial focal plane instruments. (NASA illustration MSFC-75-SA-4105-2C, http://mix.msfc.nasa.gov/abstracts.php?p=2585.)

requirements of the project and a scientific advisory committee to define the scientific requirements.[6]

Concerned with the gap between the aerospace industry vision and what was acceptable to astronomers, I decided that my role was to foster more realistic studies of the feasibility of a space telescope that would be acceptable to as many of the stakeholders as possible. Therefore, I organized a study committee by bringing together active astronomers and NASA engineers. I got them to sit down and develop the specifications for a large, presumably three-meter, space telescope that would satisfy astronomers' requirements and still meet the feasibility concerns of the engineers. This step, I feel, was really the birth of what ultimately became the HST (Figure 1.1-2).

Of course, there were many pressures, from many directions, defining what would eventually become the HST. I vividly recall the scaling down of the mirror size and the lingering question of how this would reduce its value to astronomy. The Marshall Space Flight Center was eventually assigned the responsibility for turning our sketches into a design. Robert O'Dell was hired as the project scientist with the detailed responsibility for keeping the scientific requirements at the center of the planning. I maintained a general overview of the continued developments as program scientist, and we both repeatedly dealt with the size of the telescope.

There had been strong pushes to decrease the diameter of the mirror, probably to make use of facilities that existed for other purposes. We were asked to consider mirror sizes of 2.4 meters and even 1.8 meters. We resisted by citing the requirements that studies by astronomers had laboriously developed.

For instance, a primary objective of the telescope was to determine the brightness of Cepheid variables in the Virgo cluster of galaxies. Edwin Hubble had shown that the velocity of recession of distant galaxies was proportional to their distance. However, the proportionality constant was uncertain by a factor of two. Galaxies have random motions. The random velocities of distant galaxies are small compared with the velocity caused by expansion, but for nearby galaxies these random motions overwhelm the general expansion. Moreover, the nearby galaxies are in a group in which they interact gravitationally. To determine the proportionality constant it was necessary to determine the distance of a cluster of galaxies not interacting with nearby galaxies and distant enough that the random velocities are not significant on the average. The nearest suitable cluster is the Virgo cluster of galaxies at a distance of about 54 million light years. Henrietta Leavitt had shown that the brightnesses of a particular class of variable stars, called Cepheids, were an accurate function of the periods of variation. We could calibrate this relation for Cepheids in the Milky Way galaxy. Thus if we could observe these variables in the Virgo cluster, we could determine the distance of the cluster. Measuring the velocity of the expansion was easy. I and, independently, Bob O'Dell and several others determined that with the available detectors, we could reach the Cepheid variables in the Virgo cluster with a 2.4-meter mirror but not with a 1.8-meter mirror. Dropping the mirror diameter to 2.4 meters also made easier the design of a satellite that would fit the Space Shuttle. As the early design developed, it was necessary to make a place for the project in NASA priorities. It was relatively easy to convince my superiors in NASA that such a telescope would be worth the cost. Convincing the political community, with little understanding of science, was more difficult. James Webb, the administrator of NASA at that time, gave a series of dinners for men with political power. After each dinner, three of us presented a "dog and pony show." Jesse Mitchell discussed the engineering and its feasibility, Richard Halpern presented the management plans, and I described the scientific research we expected to do with the telescope. I never testified before the U.S. Congress, but I did write congressional testimony to justify the LST for about 10 years. I also pitched the case for the telescope to representatives of the Bureau of the Budget (now the Office of Management and Budget), the agency that prepares the budget the President sends to Congress. At some point, for political reasons, the word large was dropped from the name, and the satellite simply became the Space Telescope until launch.

But the political community was keenly aware that not all astronomers wanted the LST. So convincing politicians was a far different problem indeed. By 1974, after several years of effort by advocates, the Congress seemed completely uninterested in hearing about the prospects of the LST. At this point a few astronomers, primarily Spitzer at Princeton University and Bahcall at Princeton's Institute for Advanced Study, rallied their colleagues nationwide to lobby for the LST.

First of all, Spitzer, Bahcall, and colleagues had to counter a major criticism that had been made by skeptical members of Congress. Early decadal surveys conducted by astronomers and convened by the NAS as to what projects to support in the 1960s and 1970s had not highlighted the need for the LST. Some thought the results of the decadal surveys were due to a selective process that did not represent the true needs of the entire astronomical community; this indifference to the LST had to be answered before Congress would act. Accordingly, the then-called Space Sciences Board of the NAS undertook an additional study that elevated the LST to a top priority. This influenced Congress to relent; NASA received half of the funding that had been requested for the LST, and it became a "new start" project in fiscal year 1978.[7]

After construction was started, however, Congress cut the budget below an optimum level. Of course, this cut in funding increased the final cost of the mission. Only through major lobbying efforts mounted by astronomers starting in the 1970s and continuing into the 1980s did Congress approve the project, belatedly, at full funding.

There were many instances along the way that make me wonder today how it all ended so well. At one point Senator Proxmire, noted for ridiculing government funding that he considered frivolous, asked NASA why the American taxpayer should support an expensive telescope. I did a back-of-envelope calculation and determined that for the cost of one night at the movies for all Americans, every American would enjoy 15 years of exciting discoveries from the HST. I was probably off by a factor of four or five, depending on how launch and servicing costs are allocated, but we shall probably have 25 years of discoveries. Even at a cost of a night at the movies once a year, which would more than cover costs by any accounting, I believe that most Americans believe that the expenditure has been worth it.

At the time the HST was being designed, NASA was pitching the Space Shuttle as a cheap way to launch spacecraft. To keep costs low, however, NASA needed to launch many shuttles and so was hungry for payloads. Therefore, NASA directed that all satellites be designed for launch by the shuttle, and several were designed to be serviceable. The HST was scheduled to be launched by the flight following the *Challenger* accident. That catastrophe cancelled all shuttle launches for three years, during which time the satellite was kept in storage and a critical number of knowledgeable engineers kept on the payroll until what became the 1990 launch.

These three wasted years added significantly to the cost of the mission. The *Challenger* experience caused NASA to rethink its use of the shuttle for most missions. Most payloads had to be redesigned for robotic launches. Fortunately, the HST was too far along to be changed. The ability to service it with the shuttle not only saved the basic mission after the mirror problem was discovered but also provided the possibility of replacing instruments from time to time by more modern versions, thus greatly increasing the capability of the telescope (Figure 1.1-2).

I took advantage of an early out period to retire in 1979 but continued for nine months longer as program scientist to participate on the Source Selection Board, a body that was charged to select the contractor for the Space Telescope Science Institute (STScI). I found this an interesting experience. There were five proposals, four of which based STScI at the Institute for Advanced Studies at Princeton. The proposals from Associated Universities Incorporated, which managed the National Radio Astronomy Observatories, and from Associated Universities for Research in Astronomy, which managed the National Optical Astronomy Observatories, were highly competitive, and the decision between them was difficult. The latter placed the STScI at Johns Hopkins University in Baltimore. Many people believed that it was selected because Baltimore is closer to Goddard Space Flight Center. That has helped over time but did not enter our deliberations.

I left the project before substantial management problems arose, leaving their solution to my successor, Ed Weiler. He also had to handle the discovery of the mirror problem. It was clear from his actions in these major fiascos that I had left the project in good hands.

Notes

1   John N. Bahcall and Jeremiah P. Ostriker, "Lyman Spitzer Jr.," *Physics Today* 50(10) (October 1997):123–24.

2   On the IGY see Roger D. Launius, James Rodger Fleming, and David H. DeVorkin, ed., *Globalizing Polar Science: Reconsidering the International Polar and Geophysical Years* (New York: Palgrave Macmillan, 2010).

3   Space Science Board, *A Review of Space Research* (Washington, D.C.: National Academy of Sciences, 1962). On the manner in which space science projects were developed see John E. Naugle, *First among Equals: The Selection of NASA Space Science Experiments* (Washington, D.C.: NASA SP-4215, 1991).

4   The Boeing Company, "A System Study of a Manned Orbital Telescope," quoted in Nancy Grace Roman, "Space-based Astronomy and Astrophysics," in *Exploring the Unk\nown: Selected Documents in the History of the U.S. Civil Space Program, Volume VI, Space and Earth Science*, ed. John M. Logsdon (Washington, D.C.: NASA SP-2004-4407, 2004), 532.

5   Gordon C. Augason, NASA Headquarters, "Manned Space Astronomy," November 1966, Space Telescope History Project, National Air and Space Museum, Smithsonian Institution, Washington, D.C.

6   Space Science Board, *Scientific Uses of the Large Space Telescope* (Washingtn, D.C.: National Academy of Sciences, 1969). Although LST stood for Large Space Telescope, in the minds of many astronomers it also stood for the Lyman Spitzer Telescope, given Spitzer's seminal role in proposing the concept.

7   Roman, "Space-based Astronomy and Astrophysics," 533–534.

2

Steps Toward the Hubble Space Telescope

C. Robert O'Dell

We would not be where we are today if Nancy Roman had not done what she did early in her career. Among her many accomplishments, she chaired the National Aeronautics and Space Administration (NASA) Headquarters advisory group in 1971–1972 during the feasibility studies (Phase A) of the Hubble Space Telescope (HST). She was joined by Marc Aucremanne from NASA Headquarters, Ernst Stuhlinger from the Marshall Space Flight Center (MSFC), Anne Underhill from the Goddard Space Flight Center (GSFC), and outside astronomers Aden Meinel, Bev Oke, Lyman Spitzer, Joe Wampler, and me. The GSFC and MSFC engineers and multiple contractors conducted feasibility studies of what a large observatory in space could be and reported to us regularly.

Now, why would astronomers want to put a telescope in orbit? Throughout the pre-telescopic period, the human eye set the angular limit of our images at about sixty seconds of arc, about one-thirtieth of the diameter of the Moon. Through the advent of the astronomical telescope with Galileo's application of an existing invention, suddenly we could see the universe in much more detail. Telescopes' images became gradually better over the next several centuries as the quality of the telescopes improved, but they were always limited by being underneath the turbulent Earth's atmosphere.

We knew, even back in 1971, that building a three-meter-class telescope in space would be as big a step as the application of the first astronomical telescope, and therefore doing so was a very attractive goal.[1] During 1971–1972 many configurations for the observatory, and how to operate it, were considered. An early concept from the GSFC showed a cross section with an astronaut present, somewhat akin to the astronomical observatory contained on Skylab (Figure 1.2-1). The concept drawing clearly showed a contaminating astronaut present in the back during a servicing mission. Contamination from an outgassing human body was deemed acceptable at that time.

At the end of the feasibility study, we entered the preliminary design

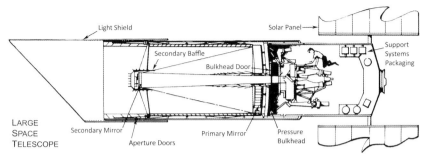

Figure 1.2-1. This early GSFC concept of the design for the HST (then the Large Space Telescope) packaged most of the heavy components at the aft end of the observatory and presumed that shirt-sleeved astronauts would be able to service a complex assembly of scientific instruments. This figure was presented to the Phase A science advisory group by Frank Cepollina of GSFC. (From the files of the author.)

phase (Phase B). The MSFC had been selected as the lead center for the project, and GSFC was responsible for the scientific instruments and the long-term operations. Parallel and competitive contracts were let for development of alternative designs for the observatory. I came on board as the first project scientist in September 1972 and relocated from the University of Chicago to NASA's MSFC in Huntsville, Alabama.[2] In December of that year, NASA Headquarters issued an announcement of opportunity for the creation of instrument definition teams that would define scientific instruments for the Large Space Telescope (LST) plus a few generalists who would help form the science working group to oversee the scientific goals incorporated into the preliminary design. As part of implementing that announcement of opportunity, a dog-and-pony show was put together by NASA people, and we went around to major ground-based astronomy centers in the USA to tell just how good a thing the LST would be.[3]

We needed a broader constituency in Phase B than in Phase A because many of the people involved in space astronomy at the time had been there from the initial conception, and they were not representative of the entire community. This broader community was obtained in part by holding special sessions at the American Astronomical Society meetings; the watershed event, however, occurred on 30 January 1974 at the American Institute of Aeronautics and Astronautics (AIAA) annual meeting. As background one has to realize that the most influential astronomer of the time was Jesse Greenstein of the California Institute of Technology because he had led the decadal study of what should be done in developing new astronomical facilities. Jesse's personal opinion was that it was better to take the roughly identified cost of $300,000,000 for the LST and build twenty more 200-inch (5-m) telescopes. This was a common view. His committee considered the LST as

Figure 1.2-2. One of the greatest challenges facing NASA was to design the procedures for deploying and then servicing the HST. This involved creating tools and ways to employ them that could be conducted by astronauts in space gear in zero-gravity. The Neutral Buoyancy Simulator (NBS) at the MSFC proved to be an effective facility to develop these practices. Here, two astronauts train at MSFC's NBS in 1985. One uses a foot restraint on the Remote Manipulator System; the other performs maintenance techniques while attached to the surface of the HST mock-up. (NASA image MSFC-75-SA-4105-2C, http://mix.msfc.nasa.gov/abstracts.php?p=1795.)

something for the future, beyond their recommendations.[4]

But at the 1974 AIAA meeting near Washington, D.C., we were able to convince major astronomers (Allan Sandage, Ivan King, Margaret Burbidge, Gerry Neugebauer, and George Herbig) who had not been involved with space astronomy before to "come out" and advocate for the LST. I consider this an important epoch for the HST program because not only did these scientists join us in advocating for the observatory but also Jesse participated and spoke favorably about the project.

During Phase B the design evolved into the final 2.4-meter instrument, which was inherently simpler to build than the original 3-meter configuration. This design was simpler because one could now fit the support equipment (e.g., batteries, reaction wheels, and gyroscopes) around the primary mirror and near the center of gravity and still fit the telescope within the shuttle payload bay. This configuration reduced the moment of inertia of the design, making it possible

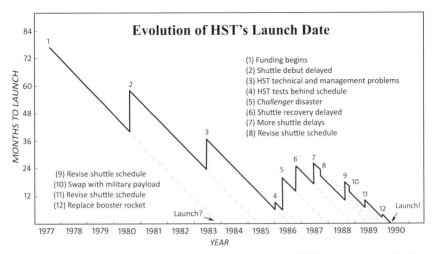

Figure 1.2-3. The launch date of the HST slipped throughout the 1980s, as documented in this graphic prepared for the 2003 HST-JWST transition hearings. (Graphic courtesy of Richard Tresch Fienberg.)

to build a simpler pointing and control system.

We also finalized the plans for maintenance and refurbishment, which was planned to occur at about three-year intervals. Many options of servicing were considered on almost a cut-and-try approach; that is, we were designing something that had never been done before. The engineers would come up with a design feature, it would be tried in the Neutral Buoyancy Simulator at the MSFC (Figure 1.2-2), and then a decision would be made whether to adopt or reject the idea.

Finally, in fiscal year 1978, we started the hardware phase of the project, which lasted longer than we expected. At the beginning, we expected a launch in 1983, but problems of cost and schedule arose repeatedly in at least two cycles. A new total budget and schedule were negotiated, and a new project manager was appointed. We were scheduled to launch in the autumn of 1986 when the *Challenger* accident occurred. Over the next several years the delays in launch were primarily driven by the availability of transportation; that is, the Space Shuttle (Figure 1.2-3). Soon after launch, the problems with the shape of the primary mirror and the vibrations caused by the solar arrays were discovered. The recovery of the original design goals through the highly successful first servicing mission will be described later in this volume. The recovery was made possible because of the flexibility that was designed into the observatory. The first servicing mission was the finest hour for people working at the Space Telescope Science Institute, MSFC, the GSFC, and the Johnson Space Center.[5]

Notes

1   Lyman Spitzer Jr., "Astronomical Advantages of an Extra-terrestrial Observatory," Project RAND 30 July 1946.

2   Nancy G. Roman, program scientist, and C. R. O'Dell, project scientist, to members of the LST Operations and Management Working Group, with attached, "Minimum Performance Specifications of the LST," 12 February 1975, Space Telescope History Project, National Air and Space Museum, Smithsonian Institution, Washington, D.C.; George B. Field, Director, Center for Astrophysics, Harvard College Observatory, to Dr. James Fletcher, NASA administrator, 12 February 1976, NASA Historical Reference Collection, History Office, NASA Headquarters, Washington, D.C.

3   NASA Headquarters, "Announcement of Opportunity for Space Telescope," March 1977, Space Telescope History Project, National Air and Space Museum, Smithsonian Institution, Washington, D.C.

4   Space Studies Board, Astronomy and Astrophysics for the 1970s, 2 volumes (Washington, D.C.: National Academy of Sciences, 1972–1973).

5   The earliest period of the HST program has been well described by Robert W. Smith, *The Space Telescope: A Study of NASA, Science, Technology, and Politics* (New York: Cambridge University Press, 1989, rev. ed. 1994); a highly personal short history appears in Chapter 10 of C. Robert O'Dell, *The Orion Nebula* (Cambridge, Mass.: Harvard University Press 2003); and a fascinating prediction of how the HST would be used was made by Malcolm Longair, *Alice and the Space Telescope* (Baltimore, Md.: Johns Hopkins University Press, 1993).

3

Building the Hubble Space Telescope as a Serviceable National Facility

Edward J. Weiler

The history of the maintenance and refurbishment (M&R) of the Hubble Space Telescope (HST) began when I (Figure 1.3-1) was a program scientist at National Aeronautics and Space Administration (NASA) Headquarters. In that context there was not much I could do other than fight for level-one performance specifications—level-one specs as we called them—and try to ensure that we delivered the telescope that we had promised to the U.S. Congress and the world. In the history of HST, M&R plays a critical role and would make a good subject for a book in and of itself because it has been quite a roller coaster ride.

Even though I did not join NASA Headquarters until 1978, my involvement with the Space Telescope actually started in 1976 when I was a newly minted Ph.D. from Northwestern University working at Princeton University. My first boss at Princeton was a still young and very energetic Lyman Spitzer, whom I did not know at the time was going to become the father of

Figure 1.3-1. Edward J. Weiler (at podium) with Nancy Grace Roman and C. Robert O'Dell on stage at the HST symposium in 2009. (Smithsonian Institution image; photo by Eric Long.)

the HST. My job was conducting research on the Copernicus Experiment Package, the 32-inch (0.8 m) ultraviolet telescope aboard the Orbiting Astronomical Observatory 3 satellite that had been launched in 1972 and was a major project Spitzer had spearheaded. In my spare time, I went to meetings that Lyman held for planning to build cameras for the Large Space Telescope.

When I joined NASA Headquarters, Space Telescope was scheduled to be only five years from launch, and it was going to cost only 400 million dollars. Those were the best of times for me—entering an arena and a mission that would be so dynamic! The telescope was planned from the very start to be serviced, and though we later did some things that we did not plan to do, the crucial point is that it was designed to be upgraded and we planned for it.

Parenthetically writing from the perspective of 2010, NASA has moved away from a philosophy of M&R and no longer designs spacecraft with it in mind. The Space Station is the only spacecraft we have of late designed to be serviced, so it is especially important to look back now to see how M&R was done. Sadly, if we returned to an M&R philosophy—to design and build scientific spacecraft that could be rescued, serviced, or improved—we would have to start from scratch.

The early M&R program for the Space Telescope was exceptionally aggressive. The refurbishment part of M&R was the most interesting part because, as both Bob O'Dell and Nancy Roman well know, back in 1978 the plan was that with the aid of the Space Shuttle we would service the telescope every two and one-half years. Then every five years we would go up with the shuttle and bring it back down to the ground, disassemble it, send it out to all the contractors to rebuild, put it all back together again, and then redeploy it once again with the shuttle. That was the plan, and it sounded good.

In the early 1980s, however, budget limitations and technical problems brought reality knocking. The telescope's cost escalated and its schedule started slipping; these factors had serious implications for M&R, including the plans for building orbital replaceable units (ORUs). These included science instruments, batteries, gyros, reaction wheels, and the like. The ORUs were critical to servicing because they were designed to be taken out of the spacecraft by astronauts working inside spacesuits with thick, heavy gloves. We designed the ORUs to be easily replaceable, putting them on rails or having attachments with just four easily disconnected bolts and blind mate connectors. This was not really high-tech stuff but very expensive to design. For the average spacecraft, matters are totally different of course: bolting an instrument in, sealing it up, wiring it, and not letting anyone touch it before launch. But the plans for the ORUs began to be cut back as the HST (as now called) budget started increasing in the early 1980s.[1]

In 1983 I was intimately involved, as was Bob O'Dell, in the debates concerning M&R, which was then under attack. It would be too costly some argued; others insisted it was critical to the success of the mission. Many believed that the one part of M&R that was never well thought out early on was how to develop replacements for the science instruments. We really did not have an aggressive plan. It was great that these instruments we were constructing were replaceable, but where was the budget to build the replacements? Where was the plan to solicit new instruments and to replace the old ones? So we generated a white paper and released it on 18 December 1983—that date is significant for reasons to be seen shortly—that proposed an aggressive science instrument replacement plan. I wish we could claim prescience back then, but it turned out that the crucial concept was that imaging might be very important to the HST. It would be important not just for the science that would be done with the cameras but also for engendering public interest in the HST. One of the first things we sought to ensure was that we could replace the main camera, the Wide Field Planetary Camera (WFPC; Figure 1.3-2).[2]

The WFPC best symbolizes the M&R philosophy. It was the only instrument, beyond the primary mirror, to have a backup under construction before any plans were afoot to replace the original since the success of the mission required that the camera work. Also, the backup was to be as close to the

Figure 1.3-2. Astronauts Jim Voss and Jay Apt in 1991, assisted by several technicians, practice routines for replacing the WFPC on HST in the water tank at Marshall Space Flight Center's Neutral Buoyancy Simulator. (NASA image MSFC-75-SA-4105-2C, http://mix.msfc.nasa.gov/abstracts.php?p=1617.)

original as possible, differing only by a degree made possible by improvements in detectors at the time it was designed. So the concept of a WFPC "clone" was proposed in December 1983. It was approved by the Science Working Group, with Bob O'Dell serving as the chair at that time. The clone got pretty strong support there. The program manager, of course, had said "no way in hell, there's no money in the budget to do this, ta-da, ta-da, ta-da!" To make a long story short, through the lobbying by John Bahcall and testimony to the right congressional people, suddenly the deputy associate administrator of NASA, Sam Keller, who was testifying a few days later, told Congress when asked, "Of course we're going to build a WFPC clone. It's very important." So that is a part of history I do not think has been publicly reported before, but John Bahcall deserves the credit for really making the WFPC clone happen.

In the mid-1980s there were still more budget problems for the HST. A study of M&R by the Marshall Space Flight Center—Marshall became the lead NASA center for the HST—prompted NASA Headquarters to change fundamentally the whole concept of M&R. The study showed that the existing plan of returning the telescope and then sending it up repeatedly with the Space Shuttle every few years would cost billions and billions of dollars. This earlier plan just did not make any sense. And this was even at a time when we did not have to count the cost of shuttle launches in our cost figures for a scientific spacecraft. We scientists paid for only the HST, not all that other good stuff to do with launches.

So in the mid-1980s, we reviewed the M&R plan, and this review led to a new plan that would see the Space Shuttle visiting the HST about every three years to service it. That was when the term service started being used, and the term M&R essentially disappeared because refurbishment was not in the cards. By this time, too, the WFPC clone had been funded, and so it was on its way. The *Challenger* disaster in 1986 meant that we had to wait until April 1990 for the HST to be launched.[3]

But we did finally launch the HST in April 1990. Remember that the originally proposed cost for the HST was 400 million dollars in 1983 dollars. It is amazing how few people remember this figure today because the HST has been such a great success. You hear all the talk about the overruns we are experiencing in NASA programs now, as if something like that had never occurred before. But the HST was launched in 1990 for a total development cost of 1.6 billion dollars. That is a 300 percent overrun, but as far as I can see, it was worth every penny of it!

So we were on top of the world in April 1990 with the HST safely in orbit. We were making all kinds of promises, great promises such as the HST was the biggest leap in astronomical capability since Galileo. Some

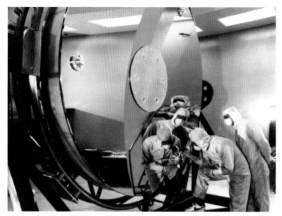

Figure 1.3-3. The 94-inch (2.4-meter) primary mirror for the Space Telescope on a transport frame in 1982, just after it was vacuum electro-coated with its reflective aluminum surface and then over-coated with a transparent layer of magnesium fluoride. (NASA image MSFC-75-SA-4105-2C, http://mix.msfc.nasa.gov/abstracts.php?p=1619.)

of the press suspected we might be hyping all those promises, but we believed in ourselves and in the HST.

Then after two months of not being able to focus the telescope's optics, on 27 June 1990, a day that shall live in infamy, we had to report to the world that the mirror (Figure 1.3-3) had spherical aberration. Yours truly got the honor to tell the world at a press conference that day about what that meant scientifically for the HST program.[4]

Lucky for me, I had some answers. On the morning of my briefing there had been a meeting of the Science Working Group, and one of the young astronomers attending, John Trauger from the Jet Propulsion Laboratory, came up to me before the afternoon press conference. John said, "You know, we seem to know what the problem is with the main Hubble mirror. It's too flat at the edges, so it's got spherical aberration, but we think we know the prescription very, very well. We built the perfect eye except it's got the wrong prescription. It's got the wrong curve in it." John then advised me that the four tiny relay mirrors in the WFPC clone that he was building, each hardly the size of a nickel, could be refigured. "If we change the figure on the surface of those mirrors, we can cancel out the error in the prescription in the Hubble primary mirror."

That sounded pretty hopeful, hopeful enough that I actually mentioned it at the press conference. But my message was not very well reported by the press. What was reported was "Hubble Trouble," there would be no images, and all sorts of stuff like that.[5] I could understand the press' reaction because the HST had not met many schedules over its long history, and here we were promising that we could fix spherical aberration with the WFPC clone. Not only were we claiming we were going to fix it by building the WFPC clone, but we were going to launch it by December 1993 and we would do all that within the budget. Nobody believed us. Stories about "Hubble Trouble" just filled the press for the next two months. The fact is, though, that the team working at Goddard, Marshall, Johnson, Kennedy, and their contractors pulled it off. We actually did wind up launching that mission by December 1993, and it was on budget.

When John Trauger had the idea of using the WFPC clone—its name of course ended up as WFPC2—to fix the spherical aberration, COSTAR was not yet even a dream. The Corrective Optics Space Telescope Axial Replacement (COSTAR) was proposed later in the summer of 1990, and detailed work on it was started up only by the end of the summer. By then we had a plan to fix not just the imaging with the WFPC2 but the light beams for the other instruments aboard the HST: two spectrographs and a European camera for which no clones had been built. A complex machine, COSTAR (Figure 1.3-4) had arms that would extend into the primary mirror's optical beam, inserting corrective mirrors right into the beam, and then sending corrected light into the two spectrographs and the European camera. That was the plan. Nobody believed we could do it, but we kept working at it.

The 2 December 1993 Space Shuttle launch, Servicing Mission (SM) 1, was called the "Miracle in Space Mission," by the Public Broadcasting Service's program *Nova*, which covered the mission.[6] We had hoped that by the time of

Figure 1.3-4. Astronaut Kathryn C. Thornton (top)—anchored to a foot restraint on the end of the Remote Manipulator System (RMS) arm in Endeavour's open bay—maneuvers the Corrective Optics Space Telescope Axial Replacement (COSTAR) into position for installation in December 1993. Thomas D. Akers is at lower left. (NASA image STS061-47-014, http://images. jsc.nasa.gov/lores/STS061-47-014.jpg.)

the American Astronomical Society meeting in January 1994 we would have the first images back from the repaired HST to demonstrate to the world that we had fixed the spherical aberration. Again, extremely uncharacteristic of the HST, we actually beat that schedule dramatically.

I remember driving home to Annapolis one night and my pager went off. It was John Trauger. He was up at the Space Telescope Science Institute in Baltimore and he said, "Ed, we think we'll get the first image in tonight around midnight or so."

So I headed up to Baltimore. The first image came down on 18 December 1993, 10 years to the day after the WFPC clone had first been proposed. The image slowly came up. It took probably just a few seconds, but it seemed like six hours for that darn image to appear on the computer screen. First we saw a little bright dot. That was good enough, because it had no junk around it, that is, no fuzz like the images that suffered from spherical aberration. But then all the other faint stars started coming up as little bright sharp dots, and then we knew we had nailed it. The WFPC2 worked great. That is a day I will never forget as long as I live.[7]

Let me note very briefly that we had subsequent servicing missions: SM2 in 1997, SM3A in 1999, SM3B in 2002, and then SM4 in 2009. All of them were 100 percent total successes. Every single EVA worked beautifully.[8]

There have been various general discussions about servicing and the future of servicing in NASA. So let me conclude with some thoughts on servicing. Of sixty operating missions in our solar system the HST is the only—and I emphasize the word only—science mission in space that is serviceable. The decision not to do more servicing missions was made all the way back in the mid to late 1980s. If you do not believe me, talk to Charlie Pellerin, head of the Astrophysics Division in NASA Headquarters. He made that decision because of two things: (1) the extreme costs of a serviceable observatory—remember those ORUs on rails and blind-mate connectors, doors that opened up, and so on—and (2) the limitations of astronomical observing from low Earth orbit.[9] We are about to celebrate the twentieth anniversary of the HST, but it has not observed for a total of 20 years. It has observed for 10 years because it is in low Earth orbit and half the sky is occulted by the Earth. Low Earth orbit is not a good place to do astronomy. That is why most of our observatories, most of NASA's missions these days, get sent way out into deep space. The Lagrangian points, more than a million miles away from Earth, are regions of gravitational stability where one actually can observe 100 percent of the time, not just 50 percent of the time. The thermal environment of low Earth orbit also is not ideal because in sunlight the temperature reaches 200 °F for 45 minutes, and then in shadow it plummets to -200 °F. That makes engineering the thermal system slightly interesting. So for thermal reasons, low Earth orbit is not a good choice either. Those are the drawbacks, what I regard as the negative stuff.

The great stuff is that the HST *was* designed to be serviced, we had a Space Shuttle, and the use of the shuttle has been a great deal for science. What do I mean by that? The HST cost a lot of money to do science, but that was not the full cost. The reason the HST was a great deal was because the science side of NASA paid for about half of it. We, and by "we" I mean the discipline of astronomy, did not have to pay for the Space Shuttle launches, of which there were six. We did not pay for the training of the astronauts. We did not pay for the facilities to train the astronauts. We did not pay for an awful lot of things done through the human spaceflight program. The true cost of the HST to the science side of NASA would have been about double if we were in today's world of full-cost accounting. It was a different world when the HST was proposed to be serviced back in the 1970s. It was a world in which the science part of NASA did not have to bear the cost of the human space flight program.

So my key point here is that all astronomers owe two debts of gratitude to the human spaceflight program: one is to the astronauts—the corps of specialists numbering in the twenties—who risked their very lives to save this telescope many times. And the other is to the human space program itself, from the monetary side. Of course, for both we owe a deep debt of gratitude to the American people, and I hope we have repaid their faith in us.

I often get upset when I hear my fellow scientists criticize the human spaceflight program. You still hear the debate among astronomers and planetary scientists about the value of human spaceflight versus robotic flight. It all depends upon how you define the interaction. Robots do not build robots, last time I checked. (At least not yet.) And robots are not operating the Mars Exploration Rovers right now; humans are. People are operating the HST from the Space Telescope Science Institute in Baltimore and the Goddard Space Flight Center. The James Webb Space Telescope will be operated by people. I do not think the critics understand that half of the cost of the HST was borne by the human spaceflight program. I have been so intimately involved with the HST for 30 years that when people talk about the separation of robotic and human, I kind of get bewildered. I mean, I would not have a career if it were not for the human spaceflight program. The HST would be an orbiting piece of space junk, frankly, if it were not for astronauts and the Space Shuttle.

But we have moved on. The Space Shuttle is no longer flying, and we are moving into a new realm. I wait anxiously to learn what the future holds.

My view stems from the fact that I do not separate science and exploration. I became a scientist because I thought that was exploration, and how you ever separate those two has always been beyond me. If we go beyond Earth orbit to places such as near-Earth objects, I can see a tremendous advantage of having human geologists walking around on those objects to figure out what they are made of—even more so on Mars.

I would like to think that in my capacity as NASA associate administrator I shared responsibility for the two Mars Rovers. I love my Mars Rovers, but they go so slowly. I cannot wait for the day that we have a human biologist with a shovel on Mars—especially an astrobiologist, who can go as far in one day walking as the rovers can go in three years. More importantly, an astrobiologist has something that can never be programmed into those rovers, and that is human intuition, scientific intuition, knowing exactly which rock to look under for that fossil or which crevice to look through. I do not know how you program that. I know there are information technology scientists out there who will say that they will build robots as smart as humans, and I am sure they will, but that is a long time away. I believe in the human mind as being the ultimate in intuition, and if we find life on Mars someday, either current life or evidence of past life, I think—and this is coming from the robotics guy—that it is going to be found by a human biologist on Mars, or a human geologist.

What will be the HST's fate? Again, we could never have predicted this, but the Sun decided to turn off at solar minimum and is barely starting to turn back on again. And why is that important? Well, high levels of solar activity send out more energetic particles, which makes Earth's atmosphere expand, which increases the drag on the HST. So our current predictions are that the HST's orbit is probably safe until the mid-2020s at least, so we have some time to prepare what we will do. I am sure by the mid-2020s we will have available Orion and other rockets. So you could think of sending astronauts up to the HST to attach a de-orbit module, and this is already being designed. Or perhaps you could even send a robot spacecraft up there that attaches some kind of retrorocket onto the HST.

The absolute bottom line in all this is we have to do something. The HST on earth weighed 25,000 pounds (~11,350 kg) and has some very large pieces. Much of the HST will probably make it through the atmosphere when it eventually re-enters. It would not be a good day if any parts that make it through the atmosphere, such as the mirror, land on a city. So we must de-orbit the HST safely into the Pacific Ocean. A retrorocket will have to be attached somehow, and that is actually in the long-term NASA budget plan. Whether humans attach the retrorocket or if that will be done robotically, we are fortunate that this is one of the few cases for which there is plenty of time to plan.

A long time ago I had the dream of placing the HST in the Smithsonian at the end of its life. I take the blame for that idea. Given that this country needs scientists and engineers, I thought it would be really neat to offer inspiration with the real HST in the National Air and Space Museum instead of having a mock-up of the HST on display. It would be great for visitors to see the actual HST, the HST that will have traveled around five billion miles by then, for kids to be able to go up and see the real thing. It is still a great idea, but I will not try to predict if we ever will be able to do it!

Notes

1   Altogether, some 70 items in the HST were ORUs. They included such items as cameras, batteries, and sensors. A full list of ORUs may be found online at "Returned Hardware: Orbital Replacement Units," http://setas-www.larc.nasa.gov/HUBBLE/HARDWARE/hubble_ORU.html (accessed 22 August 2011).

2   Ron Sheffield, "Hubble Space Telescope," 2011, http://www.sidchapters.org/ba/Archives/2011/Hubble%20Presentation%20at%202011%20Annual%20Bay-Area%20SID%20Dinner.pdf (accessed 14 July 2011).

3   NASA Management Operations Working Group for Space Astronomy, "Space Astronomy Program Plan for the 1980s and 1990s," July 1981, NASA Historical Reference Collection, History Office, NASA Headquarters, Washington, D.C.

4   NASA, "Report of the HST Strategy Panel: A Strategy for Recovery," August–October, 1990; NASA, "Report of the Task Force on the Hubble Space Telescope Servicing Mission," 21 May 1993; both documents in NASA Historical Reference Collection, History Office, NASA Headquarters, Washington, D.C.

5   Kathy Sawyer, "Hubble Expected to Yield First Images Today; Problems Encountered by Telescope Are Not Unusual among Spacecraft, Scientists Say," *Washington Post*, 20 May 1990, A6.

6   Faye Ham, "NASA Stakes Its Reputation on Fix for Hubble Space Telescope," *Science* 259 (12 February 1993):887–889.

7   Jane Opalko, "Virtual Reality Helps Out on Hubble Repair," *Odyssey* 3 (November 1994):19; James R. Asker, "Scientists Elated by Images from Refurbished Hubble," *Aviation Week & Space Technology* 140 (17 January 1994):24; John Travis, "Hubble Repair and More Wins Astronomers' Acclaim," *Science* 263 (28 January 1994):467–468; "Editorial: The Depths of Space," *Washington Post*, 2 June 2 1994, A22; C. R. O'Dell and Zheng Wen, "Post-refurbishment Mission Hubble Space Telescope Images of the Core of the Orion Nebula: Proplyds, Herbig-Haro Objects, and Measurements of a Circumstellar Disk," *Astrophysical Journal* 436 (29 November 1994):194–202.

8   Joseph N. Tatarewicz, "The Hubble Space Telescope Servicing Mission," in *From Engineering Science to Big Science: The NACA and NASA Collier Trophy Research Project Winners*, ed. Pamela E. Mack, pp. 365–396 (Washington, D.C.: NASA SP-4219, 1998).

9   Charles Pellerin, *The Great Observatories for Space Astrophysics* (NASA Astrophysics Division, 1984).

Part 2

Crisis After Launch—Restoring Hubble's Promise

Introduction:
Servicing the Telescope

Joseph N. Tatarewicz

As the first part of this volume shows, few scientific instruments—on Earth or above it—have engendered as much anticipation, hope, or indeed hype as the Hubble Space Telescope. Even before launch, its single word moniker "Hubble" had replaced in the public consciousness its full, official name or abbreviation as well as its actual, historical namesake, Edwin P. Hubble. While a small group of experts and fans knew of it in the 1970s, the general public could not escape an increasing number of pre-launch articles and features in the mid-1980s.[1] Hubble was largely assembled and tested, ready for its trip to the Kennedy Space Center (which itself promised to be a fitting spectacle), when *Challenger* lifted off in January 1986 and then met its shocking end. For more than four years, perhaps a third of its anticipated nominal life, the telescope stayed in the Sunnyvale, California, Lockheed Missiles and Space Company's clean room, while the Space Shuttle program was taken apart and reassembled. Although flights resumed late in 1986, Hubble slid further downstream to its 24 April 1990 launch on the Space Shuttle *Discovery.*[2]

While serving as curator at the National Air and Space Museum (NASM) throughout these events, and having done a postdoctoral fellowship as part of the "Space Telescope History Project" led by Robert Smith to document events in real time and produce the first book-length history, I was presented with a unique opportunity. Simultaneously, I oversaw the restoration and exhibition of Hubble's full-scale engineering test article, the so-called Structural and Dynamic Test Vehicle (SDTV), still on exhibit after nearly three decades in NASM's "Space Race" exhibition (Figure 2.0-1). The restoration of this artifact and construction of the associated display components took years, as target dates for Hubble's launch, the book's publication, and the exhibition's opening all played "tag" with one another. The joke going around Hubble's development program at the time was that if Hubble were successful, it would see not just almost to the beginning of the universe but all the way to the original launch date.[3]

Figure 2.0-1. The Hubble Space Telescope on display at NASM was created from the SDTV built by Lockheed. Since placed in orbit by the Space Shuttle in 1990 and serviced by astronauts in 1993, the Hubble has provided astronomers with a powerful tool for studying the universe. (NASM image SI 96-16364; photo by Mark Avino, Smithsonian Institution.)

The essays in this section recount the efforts to save the telescope and mission after its famously flawed primary mirror was discovered to be so, having been figured very precisely to the wrong specifications. It includes discussion of the concept of orbital servicing by the Space Shuttle, designed into the telescope from the beginning. As we in the history project divided up the research and other responsibilities, the on-orbit servicing and the Space Shuttle interfaces were my bailiwick. Since I was also at the time overseeing the restoration of the artifact, which itself had played an important role in developing procedures and tools, I shall have a little more to say about all of this below. More important than my observations, however, are the recollections of distinguished astronomers, engineers, and astronauts who actually did the work. Many of these individuals have given us valuable insight into the process, accounts of which have appeared elsewhere.[4]

Contributing to this section are four representatives of the scientific, engineering, and historical community that made the telescope a reality. Dr. John Trauger (Figure 2.0-2), senior research scientist at the Jet Propulsion Laboratory (JPL) in Pasadena, California, has been involved in one way or another with all three main cameras that have returned the aesthetically stunning and scientifically rich images for which Hubble has become so famous. Initially thinking he was just designing a replacement camera just in case something bad happened, early on referred to simply as "the clone," Dr. Trauger then found himself designing corrective optics after the mirror problem was discovered. His Wide Field Planetary Camera 2 (WFPC2), now returned from space, was exhibited for a time in NASM's "Space Race" exhi-

Figure 2.0-2. The participants in the second part of the Hubble's Legacy symposium: (L–R) John Trauger, John Grunsfeld, Harold Reitsema, and Joe Tatarewicz. (NASM image; photo by Eric Long, Smithsonian Institution.

bition, having been replaced after nearly 16 years of service in May 2009. It has since traveled to the JPL, has been on exhibit in Denver, and is awaiting reinstallation at NASM.

Dr. Harold J. Reitsema (Figure 2.0-2), an astronomer with Ball Aerospace and Technologies Corporation in Boulder, Colorado, has been involved in developing nearly all of the other scientific instruments for Hubble. Moreover, he played a key role in developing COSTAR, the Corrective Optics Space Telescope Axial Replacement, also on exhibit; an exquisite example of engineering installed with the WFPC2 on the first servicing mission in 1993. Unfolding like a child's "Transformer" toy, COSTAR inserted tiny corrective mirrors into Hubble's light path to correct the spherical aberration from the main mirror for the other scientific instruments. Without COSTAR the other instruments would have still operated with the spherically aberrated light supplied to them from the mirror in its original condition, but their efficiencies would have been painfully lowered.

Dr. John M. Grunsfeld (Figures 2.0-2, 2.0-3)—astronomer, National Aeronautics and Space Administration (NASA) astronaut, veteran of five space flights, three of them to service Hubble—has also served as NASA chief scientist. Dr. Grunsfeld is a self-professed and quite literal "Hubble hugger," who is the last human being to have directly touched the spacecraft during the final servicing mission in May 2009. All eight of his spacewalks, in total 58.5 hours, have been spent working on Hubble. If there is an extremely delicate and important telescope in trouble, he is just the person to send to work on it.

Finally, in the Appendix, Steven J. Dick, former chief historian of NASA, contributes an essay on the decision—eventually overturned—not to undertake the final Hubble servicing mission in 2010. In the aftermath of the *Columbia* Space Shuttle accident on 1 February 2003, NASA Administrator Sean O'Keefe determined that a final shuttle mission to service and extend Hubble's operational

Figure 2.0-3. Astronaut-scientist John Grunsfeld, Space Transportation System (STS)-125 mission specialist, preparing to undertake a third extravehicular activity (EVA) to service the Hubble on the fourth servicing mission on 16 May 2009. (NASA image S125-E-007926.)

life placed the lives of astronauts at too great a risk. Accordingly, O'Keefe cancelled a planned mission scheduled to take place in the latter half of the new century's first decade. Considerable deliberations led to this decision; even so the response was brutal as many people from a variety of perspectives and avenues of expertise emerged to criticize the decision. Dick's account explores this decision-making process and the response it engendered.

After the departure of Sean O'Keefe from NASA in 2005, however, his successor, Michael D. Griffin, countermanded his decision, and the servicing mission proceeded; fortunately the mission went well, and the Hubble received a new lease on life.

Each of our contributors represents many other people and teams in their areas of expertise, just as I represent a somewhat smaller corps of historians and curators. Let me explain that as historians we are taught to be critical, to maintain professional distance, to avoid getting too close to our subjects, and, above all, to avoid "going native." That being said, as an historian who has had quite a few critical things to say about NASA and its programs over my career, I need to explain that I feel no shame in revealing that I am one big Hubble hugger myself.

In the early 1970s, as seen in the first section, the Hubble and the Space Shuttle were born together. They grew up together and mutually supported one another for many years. The shuttle needed a critical mission—an important place to go and broad scientific work to do—and the telescope provided it with a signature project, one that took advantage of the shuttle's unique servicing ability. Such a huge and costly telescope, on the other hand, could not be launched and simply set adrift, as had been other smaller scientific spacecraft, and so the shuttle provided those building the telescope with the knowledge that their creation could be tended in what was expected to be an easy and economical way. It was a system, and the Hubble was to have been the first of a fleet of others, each devoted to a different wavelength and a different astronomical specialty. Other Great Observatories followed, but they were not serviceable, being placed in higher orbits to achieve full scientific value. The Chandra X-ray Observatory, launched on 23 July 1999, engaged in X-ray astronomy of the universe, concentrating on the remnants of exploded stars and even particles up to the last second before they fall into a black hole. More recently, the Compton Gamma Ray Observatory and the Spitzer Space Telescope observed the gamma-ray and infrared spectrum. Collectively the Great Observatories, led by the stunningly successful Hubble, have transformed our understanding of the cosmos.[5]

Although the actual partnership of the shuttle with a suite of astronomical spacecraft was more modest and ad hoc than the originally conceived grand vision, the shuttle–Hubble affiliation turned out to be the salvation of both, and they marched through their years together in lockstep. The irony is that today, at the end of the life of one and the beginning of the proverbial "ninth" life of the other, the system worked, and it worked beautifully. This powerful concept was proven with a genus of only one species and perhaps an eventually extinct one at that.

The Hubble's long-lived success may be credited to many scientists and engineers on the ground but also to the astronauts who flew five missions to service the instrument. Story Musgrave—and this will resonate with all who know him—told me in an interview about the first servicing mission, "I am the ballerina, and the entire ballet company's success rests upon my shoulders," articulating his deep sense of responsibility. Musgrave was quick to stress the extraordinary teamwork and coordination required, and that this sense of personal responsibility extended to all involved.

As with so many of the actual scientists, engineers, and administrators associated with Hubble, neither I nor my colleagues at the museum could have imagined that this historical effort would turn into such an intimate experience. After we identified and acquired the Hubble's full-scale SDTV, upon which the approximately 15 miles of flight wiring harnesses for the actual spacecraft were fabricated, we took great pains to exhibit it faithfully as it was employed in life to develop tools and procedures for on-orbit servicing. The display even served as a public demonstration of the servicing function when astronauts visited and were hoisted up to practice their art.

The decision to collect, preserve, and display the SDTV in NASM's "Space Race" exhibition required careful planning and considerable convincing that the effort would be worth the cost in staff and real estate. But it has proven to be a highly effective and constant reminder of the magnitude of the mission and has become the focus for the display of instruments returned from the last servicing missions as well as the iconic images gathered by those instruments and their successors. While it has all worked out rather well, we could not have imagined the twists and turns that would keep the telescope and its Earth-based simulation relevant—indeed, iconic—for three decades.[6]

Notes

1   Robert Bless, "Space Science: What's Wrong at NASA," *Issues in Science and Technology* 5 (Winter 1988–1989):67–73; "Spectrograph Expected to Reveal Images at Edge of Time," *Los Angeles Times,* 16 February 1986: Digital Collections, http://articles. latimes.com/1986-02-16/news/mn-8581_1_imaging-spectrograph, accessed 17 April 2014; Bruce Weber, "Looking Forward to Looking Back," *New York Times,* 2 April 1989: SM102 (magazine section), ProQuest Historical Newspapers: The New York Times (1851–2009), accessed 17 April 2014; Frank Morring Jr., "Excitement and Dismay at Space Telescope Center: Scientists Eager to See Farther than Ever Before, Live with Delay," *New York Times,* 19 February 1989, 51; Robert W. Smith, *The Space Telescope: A Study*

of NASA, Science, Technology, and Politics, with contributions by Paul A. Hanle, Robert Kargon, and Joseph N. Tatarewicz, rev. paperback ed. with new afterword (New York: Cambridge University Press, 1993): "Afterword," especially "Hubble Hype," p. 400. See also James Kauffman, "NASA in crisis: The space agency's public relations efforts regarding the Hubble space telescope," *Public Relations Review* 23(1):1–10.

2   Fascinating accounts of the Hubble Space Telescope may be found in R. W. Smith's *The Space Telescope: A Study of NASA, Science, Technology, and Politics*; Eric Chaisson, *The Hubble Wars: Astrophysics Meets Astropolitics in the Two Billion Dollar Struggle over the Hubble Space Telescope* (New York: Harper Collins, 1994); Tom Wilkie and Mark Rosselli, *Visions of Heaven: The Mysteries of the Universe Revealed by the Hubble Space Telescope* (New York: Hodder & Stoughton, 1999).

3   A discussion of this exhibition may be found in Martin Collins, ed., *After Sputnik: 50 Years of the Space Age* (New York: Smithsonian Books/HarperCollins, 2007), 192, 215, 220, 222.

4   Smith, *The Space Telescope*; Joseph N. Tatarewicz, "The Hubble Space Telescope Servicing Mission," in *From Engineering Science to Big Science: The NACA and NASA Collier Trophy Research Project Winners*, ed. Pamela E. Mack, pp. 365–396 (Washington, D.C.: NASA SP-4219, 1998); Joseph N. Tatarewicz, "Writing the History of Space Science and Technology: Multiple Audiences with Divergent Goals and Standards," in *The Historiography of Contemporary Science and Technology*, ed. Thomas Söderqvist, pp. 71–90 (Amsterdam, The Netherlands: Harwood Academic Publishers, 1997); Robert W. Smith and Joseph N. Tatarewicz, "Counting on Invention: Devices and Black Boxes in Very Big Science," *Osiris (New Series)* 9 (1994):101–23.

5   George H. Rieke, *The Last of the Great Observatories: Spitzer and the Era of Faster, Better, Cheaper at NASA* (Tucson: University of Arizona Press, 2006); Wallace Tucker and Karen Tucker, *Revealing the Universe: The Making of the Chandra X-ray Observatory* (Cambridge, Mass.: Harvard University Press, 2001); W. Patrick McCray, *Giant Telescopes: Astronomical Ambition and the Promise of Technology* (Cambridge, Mass.: Harvard University Press, 2004); Charles J. Pellerin, *How NASA Builds Teams: Mission Critical Soft Skills for Scientists, Engineers, and Project Teams* (Hoboken, N.J.: John Wiley & Sons, 2009).

6   The vehicle has since been refurbished to resemble the flight vehicle with support from the Goddard Space Flight Center and Lockheed. The original SDTV remains protected underneath the multilayer insulation and other additions.

4

Constructing the Wide Field and Planetary Camera 2

John Trauger

The Wide Field and Planetary Camera 2 (WFPC2) was conceived as a replacement for the Wide Field/Planetary Camera (WFPC), the main imaging camera that was launched aboard the Hubble Space Telescope (Hubble). It would be held in readiness to guarantee that Hubble's wide field imaging capabilities would not be lost in case of a WFPC failure. The WFPC was one of the six original science instruments packed into Hubble's instrument bays, which also included the European Faint Object Camera, two spectrographs, a high-speed photometer, and fine guidance sensors that provided the ability to do astrometry.

As principal investigator for the WFPC2, I am one of many individuals who took part in the development of Hubble at the Jet Propulsion Laboratory (JPL), the Goddard Space Flight Center, and across the National Aeronautics and Space Administration (NASA). We had great expectations. Hubble would be above the Earth's atmosphere, where it would be free of atmospheric turbulence. With ten times the acuity than is possible over wide fields of view from ground-based observatories, we knew in advance that Hubble would reveal familiar details in astronomical objects ten times more distant than was possible from the Earth, effectively extending our reach to a thousand-fold greater volume of space. And we would see ultraviolet wavelengths that could not be seen from the Earth. This was a once-in-a-lifetime opportunity for discovery.

With the design team and scientists anticipating the scientific opportunities yet aware of the unforgiving space environment, Hubble was painstakingly designed and assembled during years of development on the ground. Once set in motion and orbiting the Earth at 17,500 miles an hour, Hubble would have to be looked after and tended continuously, and any further refinements and maintenance would have to be performed by astronauts. Many expected there might be difficulties early in the mission, and everyone had his or her own idea of what might go wrong. Possibly, a communication antenna might get tangled with its cable or a solar panel fail to deploy. So many of the critical elements of the Hubble were designed to be serviced or replaced if necessary by astronauts.

But no one anticipated the infamous spherical aberration of Hubble's primary mirror. The primary mirror was a permanent part of the telescope and not something that could be altered or replaced. As Ed Weiler notes in his essay, the discovery of the problem began three very difficult years for us. In the news media, the name *Hubble* had, overnight, become a metaphor for spectacular technical failure (Figure 2.4-1).

While the question of how the error in figuring the mirror could have happened at all was being raised, we focused on the possibilities for correction, heartened by the support of NASA's associate administrator for Space Science, Lennard Fisk, who declared that the measure of our agency was how rapidly we could recover from problems like this. But what, specifically, were we going to do? Hubble was already designed for servicing, and so we had options. It was fortunate that we had begun work on WFPC2 in 1985, anticipating at the time that the first servicing visit by the Space Shuttle would occur three years after the launch of Hubble, then scheduled for 1986.

We had gotten the very first calibration images in early June. There was drama at the Goddard Space Flight Center as the first images of a field of stars arrived and were displayed in real time. At first we had very little time to actually examine the images, but some folks suspected that there was something not quite right about them. An early glimmer of clarity came

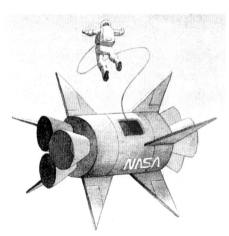

Figure 2.4-1. Which way to go? This cartoon of a rocket with engines on both ends was drawn at the time that NASA was considering its options to fix the Hubble and recover the promised science mission. (Image courtesy of John Trauger.)

when Aden and Marjorie Meinel visited our JPL laboratory to take a closer look at those first Hubble images. As the first director of the Kitt Peak Observatory and the founding director of the Optical Sciences Center at the University of Arizona, Aden had a lifetime of experience in such matters. He suggested, after a few minutes inspection, that this "looks like spherical aberration." We knew that the Hubble program folks were exploring every kind of adjustment and alignment that might bring Hubble into

Figure 2.4-2. The best fix for the spherical aberration seemed to be four small mirrors on the WFPC2, each only modestly larger than a penny. (Image courtesy of John Trauger.)

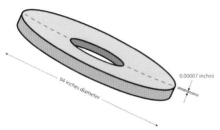

94 inches diameter

0.00007 inches

Figure 2.4-3. Diagram of the flawed primary mirror, which can be envisioned as a shallow dish that is somewhat (0.00007 inches!) shallower than intended. (Image courtesy of John Trauger.)

focus, but the Meinels provided the first suggestion of spherical aberration that I heard (see Figure 2.4-3). A few days later, Aden Meinel continued the discussion by suggesting, "By the way, you have the perfect opportunity to fix the problem with WFPC2." Small mirrors within our instrument, only a bit larger than a penny, were already designed to receive a sharp image of the Hubble primary mirror. So the opportunity, very simply, was to reshape our mirrors into corrective optics, by mimicking—in reverse—the error in Hubble's primary mirror (Figures 2.4-2, 2.4-3). At this stage in the design of our instrument we understood where our corrections would be made, but we did not know exactly what the optical prescription would be.

Pulling out all the stops, NASA was working to determine the Hubble primary mirror error by looking at a star image through various telescope focal and field positions to retrieve the prescription. Simultaneously, researchers discovered on the ground that the metrology fixture that had been used to measure the shape of the primary mirror in the laboratory had been misaligned, and their estimates were in concert with the results coming from the Hubble images.

We quickly learned that Hubble's large primary mirror had been very accurately made, but its surface had been figured to the wrong prescription. Eighteen independent analyses of the available data concluded that the shape of the mirror surface, as specified by its conic constant, was measured to be -1.0138 rather than the intended -1.0023. What did this mean? The actual shape deviated from the intended design by a very simple mathematical form, proceeding from center to edge as radius to the fourth power. The surface of the mirror departed from what it was supposed to be by only 0.00007 inches, far less than the thickness of a piece of paper. The corrective mirrors in WFPC2 optics were polished to a prescription that fit both the analysis of the Hubble images and the so-called fossil data from the metrology fixtures.

While it was important that the mirror prescription be correct, it was also important that the correction be centered on the error. So we added a tilt adjustment to the pickoff mirror that sends the telescope's light into our instrument, and three more compact tilt mechanisms within the instrument—all new features that the WFPC did not have. These tilts would provide assurance that the correction of our mirrors would be centered on the error of the telescope's primary mirror (Figure 2.4-4). If miscentered, then our correction would create new aberrations such as coma, and we would have traded one aberration for another—a foul ball rather than a home run, so to speak. We faced a situation in which we were building an instrument that fundamentally could not be focused, exactly the same way Hubble could not be focused. We had to convince ourselves and everyone else that it was going to work. At the time we delivered WFPC2 in May 1993, we had successfully completed more than a dozen interlocking tests. Any glitches would have shown up as something that we should look into further.

The astronaut servicing crew visited JPL on a number of occasions to become familiar with the instrument they would be installing during the Hubble servicing mission. We were very concerned, and the astronauts were as well, that if the pickoff mirror—indeed a very vulnerable part of a 600-pound instrument—were to be bumped or damaged at any time we would lose our mission. By the time of the Hubble servicing mission launch in December 1993, the astronaut crew had become thoroughly familiar with the WFPC2 and all the other items they were scheduled to upgrade and replace.

It was a singular time in our efforts to rescue Hubble during those few weeks between the December 1993 launch of the shuttle mission and the commissioning of the new WFPC2 on orbit. While watching the launch, and knowing the violence that our instrument was experiencing and how delicate it all was, we knew it was out of our hands. Our work was completed, and we had done everything that we possibly could to make it right. We had

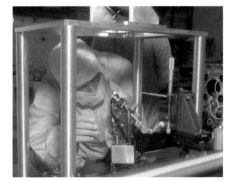

Figure 2.4-4. Four critical mirrors were given articulated mountings for on-orbit adjustment, guaranteeing that the corrective optics could be properly centered on the error on the mirror. Here a technician checks the pickoff mirror. (Image courtesy of John Trauger.)

time then to contemplate what would happen if we had not gotten it right—it would have been not just a personal disaster but a really huge setback for space astronomy. We just had to wait and see, and as Ed Weiler has noted, we did not know until one frigid night in January 1994, near midnight, in the basement of the Space Telescope Science Institute (STScI), that we had done it.

We were elated to see that first image of one star—well corrected and as sharp as expected—and we knew then that WFPC2 would produce the science that Hubble promoters had promised. For years leading up to launch, we had been thinking in technical, or even clinical, terms about the telescope's point spread function, the detailed shape of a star image. But when we actually began to see the images from the corrected system over the next few weeks, at a level of detail and fidelity we had never seen before, suddenly the technical realities were transformed into something that we really could think of as astronomers. A galaxy known as M100, which is about 20 times farther away than the Andromeda Galaxy where Edwin Hubble first observed the distance-indicating stars known as Cepheid variables, was among the many early images. Later analysis revealed dozens of Cepheid stars in those images, quite literally extending Edwin Hubble's techniques to galaxies ten times more distant than before. There were young stellar objects and flattened disks of dust surrounding young stars in the Orion Nebula, with the dimensions and appearance of our early solar system, right there in our view. We captured Supernova 1987A in time to observe the evolution of its luminous rings (Figures 2.4-5, 2.4-6).

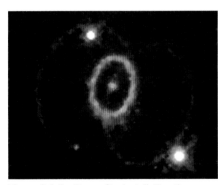

Figure 2.4-5. Soon after installation the WFPC2 was turned to Supernova 1987A, 167,000 light years away in the Large Magellanic Cloud. Scientists were fascinated by the associated ring structures and have been following their evolution, which reveals how the supernova shock waves interact with material in the vicinity of the doomed star. (NASA image; credit: Christopher Burrows, ESA/STScI and NASA, http://hubblesite.org/newscenter/archive/releases/1995/49/image/a/.)

But our goal was the recovery of Hubble's promised science, not just pretty pictures, and we needed to quickly collect the evidence that would convince the science community and various skeptics that the "Hubble troubles were over," as U.S. Senator Barbara Mikulski would declare a few weeks later. In those days, images for public release were in the form of photographic prints. For weeks the basement darkroom at the STScI was busy producing 1,500 copies of each of our selected images. Finally we had a stack

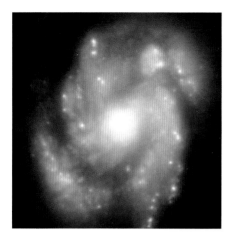

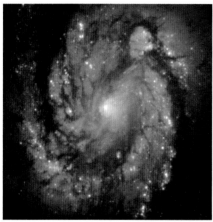

Figure 2.4-6. These before (top) and after (bottom) images of the spiral galaxy M100 were among the most reproduced image pairs illustrating the improvements to Hubble's cameras. (NASA, STScI images, http://hubblesite.org/gallery/album/galaxy/pr1994001a/npp/32/.)

of images that illustrated beyond doubt that the Hubble science mission would go forward as originally planned. Although we had the proof, we observed radio silence until all the evidence was in hand. I recall going home for Christmas with pictures I could show my wife and our children. We could say, "It's okay. It's okay." But the news was to be saved for the public announcement at the meeting of the American Astronomical Society on 13 January 1994, and thankfully it was excellent news.

We were very happy and lucky to be ready just in time for the predicted Comet Shoemaker–Levy impact with Jupiter in July 1994, just six months after our servicing mission (Figure 2.4-7). Hubble had suffered a safe-mode event just a week before: the aperture door had closed, and we were wondering if we would miss this once-in-a-lifetime opportunity. Fortunately, Hubble came out of safe mode just in time, and we captured the appearance of the impact plume within minutes of the impact itself.

The WFPC2 with its corrective optics has provided us a wealth of iconic images, many reproduced on the covers of mainstream media like *National Geographic* and even on the cover of a Pearl Jam album. One image in particular with which we have become familiar is the star-forming regions in the Eagle Nebula, popularly known as the Pillars of Creation (see Figure 3.10-1 in essay 10 by Kessler, this volume). It is clear that WFPC2 united science and aesthetics in the popular mind.

Perhaps one of the greatest surprises came from the decision by Robert Williams, the director of the STScI from 1993 to 1998, to invest 10 days of

Figure 2.4-7. Once the first servicing mission was completed, WFPC2 captured the crash of the fragmented Comet Shoemaker–Levy 9 into Jupiter in July 1994. This sequence of shots shows the impact of the G-fragment (at left) and evolution of the debris field over a five-day period. (NASA image PIA01263, http://photojournal.jpl.nasa.gov/catalog/PIA01263 [NASA Planets Selection / pia01263-Jupiter G lm#6DD15].)

Figure 2.4-8. The first Hubble Deep Field image to be released, this image was composed of 342 separate WFPC2 exposures obtained in December 1995. (NASA image; credit: Robert Williams and the Hubble Deep Field Team [NASA, STScI].)

his director's discretionary time to observe a region in deep space as empty and dark as possible, in the general direction of the Big Dipper in the northern sky. It would be a space devoid of galactic stars or bright galaxies, a place where one might peer to the beginning of time and space to see what is there. The resulting "Hubble Deep Field" (Figure 2.4-8) revealed 4,000 objects—a whole menagerie of galaxies, extending billions of light years away—all in one little area of sky about as big as a grain of sand held at arm's length. In addition to the science, we were testing the WFPC2 to the limits of its observational sensitivities.

The WFPC2 operated without fail from launch in 1993 to its return from space in the final Hubble servicing mission in 2009. It is a testament to what we can achieve in space—not just the scientists and engineers who designed WFPC2, but the technicians with the golden hands who actually built the instrument, looking after every optical mount, every mechanism, and every solder joint. This instrument continued to function successfully, electronically and mechanically, for 15.5 years on orbit—84,000 orbits and 135,000 images—leading to 1,800 science programs and over 3,700 scientific papers, probably one of the most productive astronomy experiments yet flown.

5

The Corrective Optics Space Telescope Axial Replacement (COSTAR)

Harold J. Reitsema

The Hubble Space Telescope (Hubble) contains five major scientific instruments. One of those is in a radial bay, in the best focus location that provides for the highest image quality. This is where the Wide Field and Planetary Camera (WFPC) is located on the spacecraft.

The other four instruments, mounted axially well behind the primary mirror, share peripheral portions of the focal plane. They, of course, suffered from the same spherical aberration problem in the primary mirror as did the WFPC. But the WFPC had, as both Ed Weiler and John Trauger described, a backup, and so its follow-up was started well in advance of the need. Thus, when the spherical aberration was identified, it was possible to go into a partially built instrument and fix the spherical aberration by making changes to the optics. But the axial instruments were already up on orbit and were not serviceable individually. You could not get inside those existing instruments, and, of course, a single servicing mission could not replace five instruments. It could replace a couple of instruments. Therefore the axial instruments remained in need of correction. The Fine Guidance Sensors, which were used primarily to orient the spacecraft precisely, also functioned as a virtual sixth scientific instrument for astrometry. As radial instruments these would not have the benefit of corrected light after the first servicing mission. They would have to be corrected individually, as they were replaced over the lifetime of the telescope.

In addition to the Wide Field Planetary Camera 2 (WFPC2) being fabricated at the Jet Propulsion Laboratory, there were two additional second-generation instruments already being built at Ball Aerospace and Technologies Corp. (Ball), and it was possible to correct those internally, just like the WFPC2 corrected the optical aberration for itself. So all future installed instruments could be fixed optically using the same solution that John Trauger employed on the WFPC2. But the instruments already on orbit were not operational without some magic being done. The Corrective Optics Space

Telescope Axial Replacement (COSTAR) became that magic, and so I want to describe to you just a little bit about how the magic was done.

I certainly echo John Trauger's comment that all of Hubble, and COSTAR as well, is a result of an incredible team effort; this was a huge project of really brilliant minds and committed people, starting within the National Aeronautics and Space Administration (NASA) centers that were involved as well as in the Space Telescope Science Institute in Baltimore, Maryland. Everyone was helping to analyze the problem, and many contractors were engaged, of which Ball was privileged to be one.

So the question was what can we do about the three instruments that are already there? There was also a fourth installed instrument, the High Speed Photometer. Ironically, this instrument was least affected by the spherical aberration of the primary mirror and could have done its job rather well for the life of the mission. Its science potential, however, was judged slightly below that of the other instruments. Even more ironically its principal investigator, Bob Bless, had taken the precaution of building a spare axial instrument "box" called STAR—Space Telescope Axial Replacement. While the purpose of STAR was to have been a thermally and structurally benign placeholder in case an instrument needed to be taken out without a replacement handy, it wound up being the container for COSTAR and resulted in the sacrifice of Bless' instrument.

Dr. Murk Bottema, a respected fellow of the Optical Society of America and long-time Ball staffer, looked at what was being done to prepare WFPC2 and realized that by mounting optics on the Hubble internally to correct the aberration, we could also apply the same correction in front of the axial instruments and achieve the same correction without replacing the instruments. Accordingly, we could image the primary mirror on the pupil of an optical element figured so that it would take out the spherical aberration. Bottema pursued this possibility with the relish of a true convert and offered a very detailed analysis for that optical correction. He used this to convince decision makers that it was, in fact, feasible to retrofit the optical correction to those existing axial instruments.

The challenge was, though, how do you get the optics in place? Jim Crocker, who was at the Space Telescope Science Institute at the time, suggested that we extend corrective optics on stalks in front of the other instruments. These stalks needed to be stowed inside the instrument volume during installation. After placement these optics could then be extended in front of the existing axial instruments to intercept the Hubble beam and relay the corrected light into the instruments. That, then, is the genesis of the optical and mechanical solutions that are implemented in COSTAR. Crocker had his epiphany at a Hubble meeting in Germany while adjusting a typical European shower head in his hotel room. He re-enacted the Archimedean event for the Public Broadcasting Service's *Nova* documentary, *Rescue Mission in Space* (1994).

Work on the COSTAR took precedence over work on the two replacement instruments already in progress for later servicing missions. The science teams associated with the two replacement instruments wanted to accelerate their programs as fast as possible and get their new fixed instruments in place. I was taken to task by several of the scientists involved who asked, "How could Ball be so silly as to agree to get involved in this 'Hubble trouble' and take the responsibility of fixing the spherical aberration correction for the existing axial instruments?" Those who experienced the Hubble trouble know that there was a lot of stress in the system at that time, and to get involved in it, as Ball did in taking on the development of corrective optics (Figure 2.5-1), was definitely risky. But there were so many people passionately engaged in Hubble at Ball that it was a very easy decision for us to jump in despite the risk.

It took a long time for this way of thinking to process through the system, and when the contract was put in place, we had 21 months to deliver a replacement instrument. That is a challenging schedule. And there were other challenges for COSTAR. Hubble is a very, very precise optical machine—precisely wrong at first—but precise nevertheless. To get the kinds of images that later became the norm for Hubble required fantastic optical precision and exquisite optics, very precise and complex figures made even more complex in the case of the axial instruments because they are off-axis optically. It is a credit to Murk Bottema that he was able to work out those solutions. Moreover, some spectrographs work in the ultraviolet, which put increased requirements on the optical figure and cleanliness.

The deployable optical bench—the "stalks" on which the corrective mirrors would be mounted and then deployed—was a further departure from how anyone would choose to do business in space. To move an optical bench after it has been put in place by the astronauts requires optical

Figure 2.5-1. Harold J. Reitsema working on COSTAR during its development stage. (Photo courtesy of Harold Reitsema, Ball Aerospace.)

precision in these mechanical parts and, in addition, resistance to the harsh launch experience.

To take mechanisms that have to be long and slender (Figures 2.5-2, 2.5-3) and launch them safely and then be able to deploy them with optical precision is a very stressing mechanical challenge. In addition, once they were in place, they had to be turned off and they had to stay in place with optical precision. Finally, because the COSTAR deployable optical bench was so close to all those instruments employing ultraviolet optics, the lubricant for all these mechanisms, with their hinges and rolling contacts, had to be carefully crafted not to contaminate the optics.

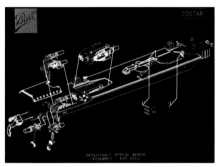

Figure 2.5-2. The key COSTAR elements. (Image courtesy of Harold Reitsema, Ball Aerospace.)

Figure 2.5-3. The COSTAR consisted of several small mirrors that intercepted the beam from the flawed mirror, fixed the defect, and relayed the corrected beam to the scientific instruments at the focus of the mirror. (Photo courtesy of Harold Reitsema, Ball Aerospace.)

In the end, we wound up with five distinct optical beams and two optical elements for each. There were twelve motors that were required to do the deployment and positioning of the optics and over 5,000 individual parts. The optics had to be folded up in a very tight bundle so that they could be inserted into the cavity in Hubble, the very small cavity behind the WFPC2 pickoff mirror and the entrance aperture of the axial instruments.

The volume into which COSTAR optics needed to be positioned was very cramped, so there were even more constraints on how the deployment could proceed. At one point—and John Trauger probably knew this—in the deployment of these optics, one of them came within a half inch of the pickoff mirror of the WFPC2.

If building COSTAR was a challenge, testing it was another and required comparable ingenuity. First, to test the correction that COSTAR promised we would need to have an input beam that looked like the

Hubble input beam in the first place. So Ball had to build a device called the refractive aberrated simulator, which produced a spherically aberrated beam just like Hubble was producing in space. Second, we had to create a simulation of the instrument bay so that we could know that when COSTAR was put in place it would be in the right location and that the deployment of the mirrors would proceed successfully without running into any obstructions. So we had to build another device that we called the Hubble optical mechanical simulator to position COSTAR just like it would be held in the axial bay structure on-board Hubble. Then, in addition, we had to demonstrate that the optical beam was corrected. So we needed to construct another simulated bay where we could put an axial instrument. Thankfully, the Faint Object Camera team had already built an engineering model that was high enough fidelity, so we mounted COSTAR next to the engineering model of the Faint Object Camera in the refractive aberrated simulator and were able to proceed with the test.

The COSTAR did not utilize a single rigid protruding pickoff mirror like John Trauger's WFPC2. Even so, we were all holding our breath during the servicing mission. Our delicate creation would now be in astronaut hands and shoved around in space manually by individuals in bulky, inflated suits. We need not have worried. The components all performed beautifully, and as a result, COSTAR performed with fantastic capability and has been a resounding success.

Once it was installed on orbit, we still needed to perform delicate alignment procedures. As John Trauger described, WFPC2 had to be aligned so that the wide field correction was on axis with the mirror's aberration. The axial instruments had to be aligned as well. So there was a process that took some time and was in sequence behind the WFPC2 to make sure that each of the channels (three axial instruments and one radial) was confocal: they all focused at the same time. The location of the optical beam had to be identified and controlled, and then the optical figure required that it be on axis.

We did have a very bad moment when the depth of the problem was revealed, but I think that the best way to describe it is to borrow a story from Charlie Pellerin, who was the Astrophysics Division director at the time. He had worked very hard for many years to get the telescope launched; upon its launch and successful orbit, he took some time off in Japan. Communications in 1990 were not quite like e-mail is now. So when he flew back to this country, he had a connection in St. Louis, and he called his boss, Len Fisk, to ask how things were with Hubble.

And Len Fisk said, "You won't believe this. Hubble's got an optical aberration, and it can't focus."

Pellerin said, "You're kidding. This can't be."

Fisk said, "Put down the phone, go out and buy a newspaper, and come back." So Charlie did that. He believed Fisk at that point.[1]

I think we all had a sense of disbelief and shock and horror. I started working with Hubble in 1977. So many people had dedicated so much of their careers to Hubble, and then to have this problem was definitely a low point for everyone. The capabilities of Hubble were so much greater than previous telescopes that all of us who had been involved in building it eagerly anticipated the first images that would be released following the April launch. Any complex satellite system requires some days of time for checkout of the health of its subsystems and commissioning of its scientific instruments, but as days passed without the release of the first images, we began to wonder what might be delaying the process. We soon heard that flight engineers were still "focusing" the telescope, a necessary task but not one that would ordinarily take days that stretched into weeks. It was not until late June that people outside NASA Headquarters and the Space Telescope Science Institute learned that there was no focus position that produced a good image. We did not immediately know that the problem was due to a correctable spherical aberration: at that point we hit an emotional nadir because we realized that the telescope had a serious optical flaw that might render it just an expensive piece of space junk.

Then what is the high point? By the time COSTAR's optical bench was to be deployed we already knew that the optical fix was successful in the WFPC2. So the unique hurdle that COSTAR had to surmount was to actually prove that the mechanism still worked after launch, and that the little mirrors, on their stalks, did in fact clear everything they needed to clear. So when we were able to extend the optics and deploy those mirrors successfully

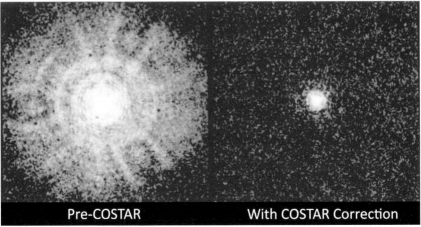

Pre-COSTAR · With COSTAR Correction

Figure 2.5-4. The imagery from the Hubble gained an order of magnitude in quality after the COSTAR fix. (Image courtesy of Harold Reitsema, Ball Aerospace.)

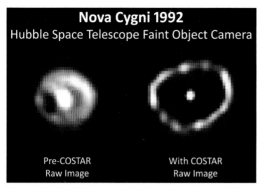

Nova Cygni 1992
Hubble Space Telescope Faint Object Camera

Pre-COSTAR
Raw Image

With COSTAR
Raw Image

Figure 2.5-5. This image also shows a before and after of the COSTAR improvement in Hubble imagery, in this figure with the Faint Object Camera photographing the Nova Cygni 1992 galaxy. (Image courtesy of Harold Reitsema, Ball Aerospace.)

we knew we were home free. It was a wonderful celebration at Ball.

The COSTAR is not a science instrument, and the instruments it corrected are not wide-field imaging devices. Its output was not the stunning pictures that have become so identified with Hubble. But one pair of images tells the tale: the star field image that we took with the Faint Object Camera before the COSTAR correction, and the corrected image obtained with COSTAR (Figure 2.5-4). You can see the absolutely spectacular improvement that was produced by COSTAR. It was the image in Figure 2.5-4 that Aden and Marjorie Meinel (see essay 4 by Trauger, this volume) and others were originally looking at; by analyzing the distribution of brightness in this star field image, they were able to deduce what the correction needed to be. This pair of images and the pair in Figure 2.5-5 of the Nova Cygni 1992 galaxy demonstrate that it was done very well.

In summary, I quote Murk Bottema, who passed away in 1992. Sometime earlier that year Ball Aerospace interviewed Murk, asking why he thought Ball had been chosen for the task:

> Well, there were actually three reasons. First of all, Ball was the only company already developing the second-generation instruments, and, of course, COSTAR can be perceived as a second-generation instrument.... [Second,] Ball had built the GHRS [Goddard High Resolution Spectrograph, also an axial instrument], and all the technology, the integration facilities, the test facilities, and the know-how could be directly applied to COSTAR. Of course, the third factor is that I invented the optics.[2]

Notes

1   Recounted by Charles J. Pellerin in chapter 1 of his book *How NASA Builds Teams: Mission Critical Soft Skills for Scientists, Engineers, and Project Teams* (Hoboken, N.J.: John Wiley & Sons, 2009).

2   Video interview of Murk Bottema circa 1992 (Boulder, Colorado: Ball Aerospace), copy in author's files.

6

Hubble: Mission Impossible

John M. Grunsfeld

The story of the rescue, yes, rescue, of the Hubble Space Telescope (Hubble) is much too farfetched for National Aeronautics and Space Administration (NASA) Public Affairs ever to have come up with it. It is just too incredible. Accordingly, I have titled my contribution, "Hubble: Mission Impossible." But the real message is, as other contributors have already commented, the Wide Field Planetary Camera 2 (WFPC2) fix, the Corrective Optics Space Telescope Axial Replacement (COSTAR) fix, and the really wonderful tools that we have that even allow work on components not designed for on-orbit repair all mean that Hubble is a story about people who do not give up. It is really about "mission possible" and about people who say, "I do think I can do that."

For the first servicing mission in 1993, because of the spherical aberration and all the attention focused on NASA, 19 review teams were looking over many shoulders. On the most recent servicing mission, May 2009, we had only one independent review team. Even so, our review team told us, pretty close to flight, that our mission was also impossible and that we should lop off a few tasks and simplify the timeline to increase our probability of success. The team that had pulled off the four previous servicing missions and my team said, "No, we can do this." Amazingly, we did pull it off. So, "mission possible" is really the theme.

That first servicing mission put NASA back on the map, restored public confidence, and demonstrated our ability to do things like build the International Space Station. In fact, virtually every single space station assembly mission space walk carries a Hubble tool.[1] So that first servicing mission, which included COSTAR, the WFPC2, and also some of the "routine" orbital replaceable units such as gyroscopes and solar arrays, set the stage. This set the bar pretty high for subsequent servicing missions (Figure 2.6-1).

The story of Hubble has never been simple or easy. The second servicing mission brought up an infrared camera, the *Near Infrared Camera and*

Figure 2.6-1. Astronaut John M. Grunsfeld, payload commander, is anchored by a handrail on the Systems Support Module while replacing one of Hubble's two second-generation solar arrays in 2002. Above Grunsfeld is astronaut Richard M. Linnehan, mission specialist, whose legs are visible on a foot restraint at the end of the Space Shuttle *Columbia*'s Remote Manipulator System (RMS). They are installing a new, third-generation solar array known as SA3. (NASA image STS109-715-038, http://spaceflight.nasa.gov/gallery/images/shuttle/sts-109/html/sts109-715-038.html.)

Multi-Object Spectrometer (NICMOS), and the Space Telescope Imaging Spectrograph. We then became even more ambitious after we got a green light to do six space walks to put up two new cooling systems, new instruments, and new solar arrays and to upgrade the Hubble computer from an i386 to an i486 on the third mission. Keep in mind, this third mission was going to fly in about year 2000, so while that processor already seemed a little bit pokey, the four i486 computers on Hubble are actually doing great. We also upgraded the data storage system from reel-to-reel tape recorders to solid-state drives.

During this time, however, the aging gyroscope system on Hubble failed, leaving at first the minimum number of gyros; then the number dropped

Figure 2.6-2. Astronaut Claude Nicollier, mission specialist from the European Space Agency, smiles for the camera as he employs a power tool to secure a storage enclosure. Image taken during STS-103 extravehicular activities (EVAs) aboard *Discovery*, 19–27 December 1999. (NASA image STS103-731-017, http://www.spacetelescope.org/static/archives/images/screen/sts103_731_017.jpg.)

below the minimum required to do science. So Hubble went dark for a while, and NASA Headquarters pulled out its "get out of jail free" card, which was to say to the shuttle program, "We have this policy that says if Hubble is in trouble, we can ask you for a call-up mission to go fix Hubble."

So we split servicing mission 3 into two, called 3A and 3B. In the first, four space walks would be required just to get Hubble back in service with a minimum upgrade (Figure 2.6-2). This included new Rate Sensor Units (which had been the cause of Hubble going into safe mode); a Fine Guidance Sensor; the i486 computer; and a few "little things," like an S-band communications transponder to replace one that had failed.

I remember it was raining quite hard in December 1999 as we went to the launch pad. We had already been delayed several months due to the shuttle's main engine problems on the previous launch. As a result we were starting to push up against the end of the millennium, and there was widespread concern about potential effects by the Y2K (millennium) computer bug. So I was particularly disappointed when we came back from the launch pad after scrubbing this launch, because it looked like we were going to slip into the next year with Hubble at risk a bit longer. To avert that, NASA leaders decided to knock off one EVA and try again to launch on 17 December.

At this next attempt I keenly recall, when we climbed into the vehicle again out at the launch pad, that despite the fine weather at the time, my launch and entry suit were still wet from the hard rains that had aborted the launch before. Finally, we did, in fact, make it up to Hubble after launch on 19 December, and we did all of that work. The single-access S-band transponder was one of those devices that were never intended to be changed out, so we had to develop new and simple tools for that procedure. It was a little bit harder than we expected. In that big, bulky space suit, I was working with subminiature assembly connectors and applying torques that were measured in inch ounces, seven to nine inch ounces. So we demonstrated that we could do a lot more than just take big things out and put little things in. We also managed to avoid problems from the so-called Y2K "disaster" that affected the rest of the planet.

Figure 2.6-3. Hubble's power control unit (PCU) was replaced by astronauts John Grunsfeld and Richard Linnehan on 6 March 2002 during the third space walk of the STS-109 mission aboard *Columbia*. (NASA image STS109-322-025, http://spaceflight.nasa.gov/gallery/images/ shuttle/sts-109/html/sts109-322-025.html.)

It was only after launch that I learned the shuttle itself was certified for Y2K but not the ground support equipment at the contingency landing site in California. The *Orbiter* could well have been powered up overnight on 31 December 1999, but with ground support equipment that was not certified! I do not know why we did not think about turning the *Orbiter* or the ground support equipment off and then turning it back on the next morning. But we were fortunate.

I was lucky enough to be leading the space-walking team for servicing mission 3B in March 2002 (Figures 2.6-1; 2.6-3), and we faced another problem—this time in the power control unit (PCU). This was a main junction box with numerous cables and wires going in and out. In preparation for this mission a simulator was built so that every night before going home I could practice taking all those connectors off, putting them back on, and swapping that box over and over again until every pin on every connector on every wire became my friend.

We had to develop very complicated and high-leverage tools for replacing the PCU because it involved linking more connectors in a shorter time than we had ever tried before. Also, with all those connectors on the same side of the box, the geometry of the access doors, and the large space suit helmet, I could not get my head in position to have stereoscopic vision to be able to align the connectors. I had to learn those details and use what became my favorite Hubble tool, which became an extension of my human fingers. I practiced the tasks with this new tool an extraordinary number of times.

I think it is easy to forget that when there is one failure, there are a thousand successes, and the people who are quietly and carefully doing the right thing and not causing any difficulty, well, we tend to forget them unfortunately. But there are a lot of people who are really dedicated to getting all those details right, and it is amazing that such a large team can function in that way and get so many things right. That is what is phenomenal to me.

During lead-up to servicing mission 3B, one of my more personal moments with NASA Administrator Dan Goldin occurred when he called the space-walking team up to his office in Washington. I thought, "Well, this is a reasonable thing. He wants to have eye-to-eye contact with the team that is going to change out this PCU." The administrator had also had a discussion with Ed Weiler, another Hubble hugger, who had to make the call as to whether we were going to change out this box. This is just one of those odd things that points to the fact that working in space is very hard. The PCU had a grounding strap that was connected to a bus bar with a bolt, washers, and a nut. Something to do with the washers and bolt had been creating some electrical resistance—measured in just tens of milliohms—between the cable and the bus bar. This was tiny resistance, but over a number of years it eventually would have caused the observatory to fail. However, if we tried to replace it and failed, we would lose the observatory right away. So there was a fair amount of pressure in making that kind of decision. Ed Weiler went to the administrator and suggested that we change it out to give Hubble the opportunity for a successful, longer scientific lifetime. The administrator went with that but with the caveat that he wanted to look us in the eye first. He said, and I quote, "I don't mean to put a turd in your punchbowl, but for the sake of the agency, you have to fix this."

Now I must say with considerable irony that prior to the administrator uttering those blunt but colorful words there was no real pressure at all for us to successfully fix the world's most fantastic scientific instrument. So I was really glad we had that meeting and had come to a common understanding! There was always great pressure to succeed. But that is when people perform their best—when they have a high-performance challenge. It is not the pressure itself but rather the challenge of building tools and figuring out how to do something under so many constraints.

If that was not enough, there was also the matter of NICMOS. The Near Infrared Camera Multi-Object Spectrograph had a thermal short in the thermal cooler, the system that kept the detectors cold enough to detect the infrared radiation. The short was from the dewar to the outside, and it caused the solid nitrogen to sublime faster. So another amazing team of engineers from all around the country invented a replacement cooling system that involved re-plumbing the telescope. In consequence, that made me more than a connector mechanic and an avionics mechanic; Rick Linnehan and I were now also plumbers on this mission. That was the hardest thing I have ever done on Hubble, physically. The new radiator did not quite fit, and we really struggled with that. I was just about at the point of giving up when I gave it one last push and it seated successfully. So, if you look at pictures of Hubble after that repair, on the outside of one axial bay is a new, big white radiator that brought the infrared camera back into operation. We also installed the Advanced Camera for Surveys (ACS), which was the imaging follow-on to the WFPC2. As an axial instrument, it supplemented the radial WFPC2, and then later most of the programs shifted over to that ACS. Repairing the power control unit, and handling the connectors, was a strikingly difficult task. To practice, we built the simulator to look just like Hubble in orbit, and we practiced ad infinitum.

One might say that after the demands of servicing mission 3B, the Hubble story started getting uninteresting. Not really! But the impossible became possible, and therefore routine. We had these scientific revolutions coming out of Hubble time and time again. But then came the tragic loss of *Columbia*, and NASA Administrator Sean O'Keefe's decision to cancel the final planned Hubble servicing mission 4. When that announcement was made in the aftermath of the *Columbia* accident, I was on loan from the Johnson Space Center to NASA Headquarters as NASA chief scientist. I actually had a letter from my boss, the chief of the Astronaut Office at Johnson, to the NASA administrator saying, "You can have John until he has to come back and start training for servicing mission 4." So I was, quite frankly, stunned when I was in a meeting with Ed Weiler and the administrator announced we were not going back to work on Hubble. We suggested to him that this might be an unpopular decision, but, nevertheless, NASA went forward with it. Reluctantly I had to support this decision despite not agreeing with it. To add injury to insult, I was tasked with describing this decision to the public as somebody who knew the shuttle, who knew Hubble, and who cared about Hubble. I learned then what it means to be in public service. But, fortunately, we persevered as a team, and the next administrator, Mike Griffin, put servicing mission 4 back on the manifest.[2]

By the time that servicing mission was revived, the Hubble imaging spectrograph was no longer functioning, and as we started training, we then lost the ACS due to a power system failure. Coming up with fixes for that, again, was really a tour de force of human ingenuity. Servicing mission 4 was really another Hubble makeover in the end.

We needed new power tools for this job—and not just power tools but also the power *of* tools (Figure 2.6-4). We might not have been able to do

Figure 2.6-4. Grunsfeld holds a power tool in the middeck of *Atlantis* while preparing for space walks during STS-125, the final servicing mission, in 2009. (NASA image S125-E-006621, http://spaceflight.nasa.gov/gallery/images/shuttle/sts-125/html/s125e006621.html.)

these repairs in space without really innovative new devices, and some of them were quite complex. For example, there was a new card extraction tool developed to work with the electronic cards (similar to those in a desktop personal computer) that have dangerously sharp edges. In a spacesuit, if I cut the fingers or the hand, I would lose the oxygen inside the suit and that would be the end of me. These tools were designed to protect us, making it not only possible but safe to handle the cards and to accomplish the tasks. There were also simple tools, such as a screwdriver that enables the astronaut not only to remove a screw but also to capture it at the same time. These were inspired by trips to Sears and Ace Hardware and seeing what people use on the ground and then employing some creative imagining for how we could adapt that to a space-suited astronaut.[3]

However, sometimes even the best tools were not enough. Mike Massimino, on the second space walk, tried repeatedly but unsuccessfully to back out the stripped retaining screws holding a handrail on the spectrograph. The entire repair depended on removing that handrail. Fortunately, the guys on the ground were always on top of things. They gave us a good idea: "Well, if you can't get the screws out, just rip it [the handrail] off." It takes about 60 pounds of force, so that is what Mike did.[4]

One of my mantras all through the training for these missions was, "Do not break the telescope." Every time we had a new crew to train for the servicing mission—and I have gone through this three times beginning with my first assignment for the 1999 mission as the astronomer on board—I told everyone repeatedly, "We don't want to break the telescope." We had silicone rubber-based gloves and spacesuits that would contaminate the optics, and folks pretty quickly got tired of me saying, "No, you do not want to touch there," "That's a bad place to touch," etc., over and over again.

Fortunately our training was long enough, especially on this last one—some three years—that eventually people started to appreciate that I was actually pretty serious about this, that it was a big deal, and that these subtleties were extraordinarily important. One example was the decision not to use the WFC3 pickoff mirror cover. It was a group decision, but I felt the risk of us trying to handle such an awkward cover in such close proximity to that delicate mirror was a much higher risk than trying to recover it. So we spent a lot of time thinking about that, and our mantra about "not breaking the telescope" got through. People would remember and stay away from the apertures of these delicate optical components, and we were really very careful not to kick Hubble.

That said, Hubble was built to take a certain amount of abuse in certain areas. On an approach to Hubble now, one can actually see handprints! It is no longer the pristine scientific observatory as it looks like in the "Space Race" exhibition at the National Air and Space Museum.[5] No, the real Hubble has handprints on it—the influence of humans is apparent because gloves deposit a little bit of material and the solar UV radiation then modifies that and incorporates it into a kind of space corrosion. And people occasionally chipped it with their tools, although overall we were pretty careful. Also, the handrails are all beat up from micro-meteoroid damage.

I should add that there is a lot of space junk in orbit, and we had to be mindful of that. The Air Force tracks tens of thousands of objects that are fist-sized or larger, and there is an exponential distribution of smaller items sized down to paint flecks, which are the kinds of things from other space-craft or real cosmic debris that account for those little teeny dings. But when you are out there doing the space walk, there is no point thinking about it.

In 1998 there was a close call on the Mir Space Station, and the crew was told to get into the *Soyuz* and to prepare in case Mir got hit. Ground control told them when the closest approach would be, so when that time had passed a call came up from the ground, "Did you see it?" The crossing velocity was such that, whatever the item was, it travelled at 10 miles per second. So at one second it was 10 miles away; two seconds later it was 10 miles the other way. There was no possible way to see it. Then I realized that question was the polite way of saying, "Are you still there?" This is part of the Russian culture. They do not say, "Hey, did you get hit? Are you leaking? Is it a disaster? Hey, are you still there?" In fact, we have calls like that on the shuttle as well that are really communication checks as much as anything else. But we do not really think about it, and the shuttle gets hit by some little tiny piece of some-thing on every mission. We just hope it does not happen during an EVA.

I am always amazed when we get to orbit, open the payload bay doors, and do not see bits of Hubble instruments floating around on their way out to space. I have never been in a train wreck, but when the solid rocket motors light you are going somewhere—it feels like you have been hit from behind by either a big truck or a train. The first two minutes are incredibly violent. Because of the payload configuration in my first flight on *Atlantis*, the vibra-tion was so high that my eyeballs were shaking in their sockets so much that I could not read the displays as we went through various resonances. It is just very rough.

I had a series of worst moments on my last *Atlantis* servicing mission, starting with liftoff when we had a master alarm. I thought, "Boy, this is an amazing story. We have come so far, and now something is wrong with the orbiter, and we are not going to make it to orbit." Then the next alarm 30 seconds later seemed to indicate our center main engine was dying, and I thought, "This is just getting worse. Now not only are we not going to make it to Hubble, but we are going to do a return-to-launch-site maneuver." Most astronauts believe the odds of surviving that are at best 50–50.

The ground said, "No action," which could mean either it is a bad sensor (and you do not have to worry about it) or there is nothing you can do and it is just going to happen. Fortunately it was a bad sensor. They had more insight on the ground. I wish they had told us that. So we continued to orbit. And then things got happy again, and I started playing with tools and space-suits, and we went out for the space walks.

Truly the lowest moment of my Space Shuttle career was during the last mission when I was hanging on the side of Hubble, and Drew Feustel was trying to release the A-latch, which is the center hole on the WFPC2 that allows you to remove the camera (Figure 2.6-5). The torque that we had available to us was not releasing the bolt. I thought, "I cannot believe we have come this far and that—after all the delays and getting near perfect detectors in the Wide Field Camera (WFC) 3—we are not going to be able to get the WFPC2 out, and it is now a fixture in the telescope." And there was a good chance that we were going to break the bolt, and it would stay in there forever. I noticed a true sense of disbelief, and my heart was really incredibly low as I was scurrying about getting other tools and things.

But then Drew Feustel, my space-walking partner, took the wrench and

Figure 2.6-5. Astronauts John Grunsfeld (left) and Andrew Feustel (standing on the RMS) install the WFC3 on STS-125 in December 2009. The WFC3's white thermal radiator is visible just behind Feustel and above the elliptical openings for the Fixed Head Star Trackers. This was the first of five STS-125 space walks and the first of three for Grunsfeld and Feustel together. (NASA image S125-E-007221, http://spaceflight.nasa. gov/gallery/images/shuttle/sts-125/html/s125e007221.html.)

put the extra elbow grease in, and this made the bolt rotate. That did not really make me feel any better until we turned it enough so that I saw the instrument slowly backing out and then recovered.

Certainly the high point during this mission—giving not so much a sense of disbelief but certainly internal good feelings—was when I put on the final new outer blanket layer (called NOBL 7). I had carried it on three different missions and finally could put it into practice. This was not scheduled to occur on this mission, but we had it with us just in case. I have always said that one of the keys to good project management, to teams, and to life (certainly family life) is do your job and a little bit more. So we were able to do our job and just a little bit more with that final installation. After I put that on, we had accomplished everything the review teams had said was impossible, and we had done it reasonably well and without breaking Hubble. That was important.

This has been the story of servicing Hubble from my personal point of view (Figure 2.6-6). For me, the story is beyond incredible; we really could not believe the spherical aberration at first, and then we could only admire the many women and men in the white suits riding to the rescue in the laboratories and test facilities all over America for the first servicing mission. And most incredible is the new science that came out afterward. Most of that new science, 50 percent of the science, we never predicted; once again we had to admit that we did not know about how amazing the universe is.

As to Hubble's lifetime, I think the big message to take

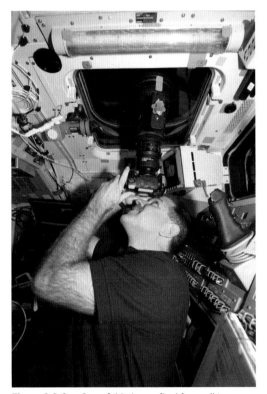

Figure 2.6-6. Grunsfeld gives a final farewell to Hubble, recording the moment with a camera peering through a window on the aft flight deck of *Atlantis*. This demanding mission had more than one glitch—but no failures. In addition to the instruments, all batteries and gyroscopes were replaced. (NASA image S125E011919, http://spaceflight.nasa.gov/gallery/images/shuttle/sts-125/html/s125e011919.html.)

away is that servicing mission 4 gave Hubble a new lease on life. Our warranty is three years, labor not included. Five years is totally reasonable. Beyond that is something we would be delighted to have, especially if it overlaps with the James Webb Space Telescope. The question of an Orion servicing mission or any other kind of servicing of Hubble is really a much bigger question about our national space priorities and where we are going to go, and that is under review right now.

The challenge now is clear. Hubble has been at work since 1990 and operating at full steam roughly since 1993. The James Webb Space Telescope is our next great telescope, and no doubt there will be amazing stories to tell about that mission, not only in building and operating it, but in the science that will come out of it. To my mind though, the question is what will we do next? Will it be serviceable or not serviceable? I think that is the grand challenge, and I am very excited and hopeful to remain a part of the outcome.

Notes

1   See Joseph N. Tatarewicz, "The Hubble Space Telescope Servicing Mission," in *From Engineering Science to Big Science: The NACA and NASA Collier Trophy Research Project Winners*, ed. Pamela E. Mack, pp. 365–396 (Washington, D.C.: NASA SP-4219, 1998).

2   See Dennis Overbye, "Scientist at Work: John Grunsfeld, Last Voyage for the Keeper of the Hubble," *New York Times*, 13 April 2009, D1.

3   See Tatarewicz, "Hubble Servicing Mission," for similar comments by Story Musgrave. Also see Tatarewicz, Oral History Interview with Bruce McCandless, Space Telescope History Project, 1984 (National Air and Space Museum, Smithsonian Institution, Washington, D.C.) on using and adapting common hardware items.

4   See Mike Williams, "Show of Strength: Rice Professor Applies Elbow Grease to Free Hubble Handrail," Rice University News Feature, 22 May 2009, http://www.media.rice.edu/media/NewsBot.asp?MODE=VIEW&ID=12611 (accessed 22 June 2010).

5   On exhibit at the National Air and Space Museum in Washington, D.C., is the original Structural Dynamic Test Vehicle, clad in thermal blanketing to simulate the flight object.

Part 3

The Impact
of Hubble

Introduction: The Impact of the Hubble Space Telescope

Steven J. Dick

In asking what the Hubble Space Telescope (Hubble) has revealed about the universe and ourselves, we are really asking about its scientific and cultural impact. As we found when the National Aeronautics and Space Administration (NASA) and the National Air and Space Museum convened a joint conference on the societal impact of spaceflight a few years ago, the subject of such impact is rich and complex.[1] One can ask, for example, what does impact mean? Who is being impacted? What is the evidence that anyone is being impacted? And if there is an impact, individuals are undoubtedly affected in different ways depending on their worldviews or individual interests and predispositions. Another way of approaching the subject in a more global sense is to ask the counterfactual question, where would we be today had there been no Hubble Space Telescope?

All four of the authors in this section illustrate how the popular impact of Hubble is intertwined with the scientific impact. While world-class science is clearly the primary purpose of Hubble, the strong popular interest has been continually reaffirmed through two decades and was highlighted by the public outcry when NASA Administrator Sean O'Keefe cancelled the fifth servicing mission in 2004 only to have it restored later by the succeeding NASA Administrator Michael D. Griffin. As Ken Sembach, the head of the Hubble Mission Office at the Space Telescope Science Institute, relates in his essay, Griffin's action extended Hubble's lifetime by many years, continuing the spectacular results to which scientists and the public have become accustomed.[2]

Most of Hubble's popular impact is undoubtedly due to its imagery, and Zolt Levay—the imaging group leader for Hubble—gives us an inside look at how these images are manipulated for aesthetic appearance while maintaining their scientific integrity. Elizabeth Kessler has also studied the aesthetics of the Hubble images and their similarity to late-nineteenth-century landscapes of the American West.[3] The importance of imagery can also be seen by comparing the imagery from Hubble with the second of NASA's

Figure 3.0-1. The Cone Nebula, a star-forming pillar of gas and dust within star cluster NGC 2264 in the constellation Monoceros, has inspired some people to suggest it as an image of Jesus. It was taken in April 2002 with Hubble's Advanced Camera for Surveys (ACS). (NASA image; credit: NASA, H. Ford [Johns Hopkins University], G. Illingworth [University of California at Santa Cruz, Lick Observatory], M. Clampin [Space Telescope Science Institute (STScI)], G. Hartig [STScI], the ACS Science Team, and European Space Agency [ESA]; http://hubblesite. org/newscenter/archive/releases/2002/11/image/b/.)

Great Observatories, the very productive but imageless Compton Gamma Ray Observatory. While there is no doubt that Compton produced world-class science with BATSE (Burst and Transient Source Experiment), EGRET (Energetic Gamma Ray Experiment Telescope), and its other instruments in terms of advancing gamma ray astronomy, most of the public has never heard of Compton precisely because its data output was not amenable to aesthetic presentation.

On the other hand, of the three Great Observatories still operating, the Chandra X-ray Observatory and the Spitzer (Infrared) Space Telescope do not seem to evoke the same reaction as Hubble, despite the striking images they produce at their respective wavelengths. This is undoubtedly due to multiple factors: Hubble was the first of the Great Observatories to return stunning images of the universe at large, it has enjoyed more than two decades of longevity in the popular imagination thanks to its unique servicing missions, and it boasts an unrivalled public relations effort. We need only recall that the famous Eagle Nebula with its "pillars of creation" (see essay 10 by Kessler, Figure 3.10-1, this volume) evoked an almost religious response in some people, as did a 2002 image of a star forming region in Monoceros called the Cone Nebula (Figure 3.0-1). Moreover, many of Hubble's other images are not far behind in their emotional impact (Figure 3.0-2). It is difficult to measure whether or not such images actually affect individual worldviews by bolstering theological convictions or simply enhancing understanding of the universe of which we are a part, just as it is difficult to measure the impact of the *Blue Marble* and *Earthrise* images from the Apollo era. But judging by their public interest and staying power, all of these images have had their impact and have enhanced the very idea of what we call "culture."

While such images are certainly evocative from an aesthetic point of view, it is their scientific content that draws us into a deeper and more intimate understanding of our universe and our place in it. The more we make the "out there" something we can comprehend, something that definitely exists apart from our efforts to understand it, the more we feel a part of it and the more it affects us. Hubble's former senior project scientist, David Leckrone, highlights some of that science in his essay and details why Hubble has been so successful. Among the factors he enumerates are not only Hubble's increase in sensitivity and resolution over a broad range of wavelengths but also its ability to evolve with technological advances through five servicing missions. As he points out, since the telescope optics were repaired in 1993 Hubble discoveries have been consistently ranked in the top tier of scientific discoveries in any given year. Those discoveries include its participation in uncovering the acceleration of the universe and the implied presence of a mysterious dark energy, confirming the existence and elucidating the nature of supermassive black holes, actually imaging protoplanetary systems known as proplyds, directly imaging extrasolar planets, and producing numerous results from the several Hubble Deep Field projects. Hubble is the example par excellence of telescopes as "engines of discovery."[4]

Hubble, as well as the other Great Observatories and spacecraft like COBE (Cosmic Background Explorer), the WMAP (Wilkinson Microwave Anisotropy Probe), and Planck all bring us to a more definitive and robust realization of our place in the universe, not only in space but also in time, in the

Figure 3.0-2. The Whirlpool Galaxy (M51) and its companion, NGC 5195, as imaged with Hubble's Advanced Camera for Surveys in January 2005. As a classic "grand design" galaxy, a term that has strangely become popular among astronomers since the 1970s, it exhibits strongly defined and articulated spiral arms that have been described poetically as a grand spiral staircase sweeping through space. The Whirlpool's arms may have resulted from a close encounter with the small galaxy NGC 5195, now situated at the outermost tip of one arm and evidently still perturbing it. Galaxy NGC 5195 has been passing behind the Whirlpool for hundreds of millions of years and is a photogenic example of what has come to be understood as an important mechanism of galaxy evolution: collisions between galaxies. (NASA image; credit: NASA, ESA, S. Beckwith [STScI], and the Hubble Heritage Team [STScI and Association of Universities for Research in Astronomy]; http://heritage.stsci.edu/2005/12a/.)

13.7 billion years of cosmic evolution. In parallel they have demonstrated in ever more detail how we originated from "star stuff," as astronomers Harlow Shapley and Carl Sagan were fond of saying.[5] In this respect Hubble and other space missions are contributing to what we might call "Genesis for the third millennium," the knowledge of our ancestry in the wake of the Big Bang. As we discern this epic of evolution, the ultimate master narrative of the universe, our deeper awareness of cosmic evolution increasingly impacts culture in numerous ways. Cosmic evolution forms the basis for the new field of Big History, which places humans in a cosmic context; it is increasingly used in educational curricula; and it is even finding a central role in the burgeoning concept of religious naturalism.[6]

Where all this will lead in the future we cannot say, but like Hubble's unexpected discoveries, I would suggest that the full impact of Hubble on our culture is as yet unknown and unpredictable. Cosmos and culture are becoming increasingly intertwined, and Hubble, through both its aesthetic images and its scientific data, is playing and will continue to play a central role in this process.

Where would we be today had the Hubble Space Telescope never existed? In short, much poorer in both science and culture.

Notes

1   Steven J. Dick and Roger D. Launius, eds., *Societal Impact of Spaceflight* (Washington, D.C.: NASA SP-2007-4801, 2007).

2   See Steven J. Dick, "Appendix: The Decision to Cancel the Hubble Space Telescope Servicing Mission 4 (and Its Reversal)," this volume.

3   Elizabeth Kessler, "The Wonders of Outer Space," in *Hubble: Imaging Space and Time*, ed. David H. DeVorkin and Robert W. Smith (Washington, D.C.: National Geographic, 2008) pp. 136–163, especially 150–151; also Kessler, *Picturing the Cosmos: Hubble Space Telescope Images and the Astronomical Sublime* (Minneapolis: University of Minnesota Press, 2012).

4   "Hubble Space Telescope's Top 10 Greatest Achievements," released on Hubble's 15th anniversary, 25 April 2005, http://hubblesite.org/newscenter/archive/releases/2005/12/background/; DeVorkin and Smith, eds., *Hubble: Imaging Space and Time*. On telescopes as engines of discovery see Robert W. Smith, "Engines of Discovery: Scientific Instruments and the History of Astronomy and Planetary Science in the United States in the Twentieth Century," *Journal of the History of Astronomy* 32(1997):49–77.

5   Joann Palmeri, "Bringing Cosmos to Culture: Harlow Shapley and the Uses of Cosmic Evolution," in *Cosmos and Culture: Cultural Evolution in a Cosmic Context*, ed. Steven J. Dick and Mark Lupisella, pp. 489–521 (NASA SP-2009-4802, 2009).

6   Steven J. Dick, "Cosmic Evolution: History, Culture and Human Destiny," in *Cosmos and Culture: Cultural Evolution in a Cosmic Context*, ed. S. J. Dick and M. L. Lupisella (NASA-SP-2009-4802, 2010), pp. 25–59. On religious naturalism see especially Ursula Goodenough, *The Sacred Depths of Nature* (Oxford, UK: Oxford University Press, 1998) and Michael Hogue, *The Promise of Religious Naturalism* (New York: Rowman & Littlefield, 2010). On Big History see especially David Christian, *'Maps of Time': An Introduction to 'Big History'* (Berkeley: University of California Press, 2004) and Fred Spier, *The Structure of Big History: From the Big Bang Until Today* (Amsterdam: Amsterdam University Press, 1996).

7

Recommissioning Hubble: Refurbished and Better than Ever

Kenneth R. Sembach

Behind every great observatory there are teams of talented, dedicated individuals who work tirelessly to bring the scientific promise of the observatory to reality. Much has been written and said about the enormous efforts it took to develop, launch, and repair the Hubble Space Telescope (Hubble), but there's a less familiar chapter of this story that is no less interesting or important. Each time Hubble has been repaired by astronauts and released from the cargo bay of the Space Shuttle back into the emptiness of space, it has undergone a several-month servicing mission observatory verification process that has readied the telescope for use. The most recent transition from repair to recommissioning occurred after the end of the stunningly successful Hubble Servicing Mission 4 (SM4) in May of 2009. For many teams, this transition marked the culmination of the successful development, delivery, and installation of hardware in the refurbished Hubble. For others, it marked the beginning of a new era of Hubble on-orbit operations and scientific discovery. The public spotlight on humans working in space shifted to the telescope and its anticipated science results.

Servicing Mission 4 resulted in the installation of two new and two repaired science instruments, bringing the total complement to six.

- *New:* The Wide Field Camera 3 (WFC3), an ultraviolet–optical–near-infrared camera with spectroscopic capabilities; the Cosmic Origins Spectrograph (COS), an ultra-sensitive spectrograph capable of resolving the wavelengths of ultraviolet light into its component colors; and a new Fine Guidance Sensor, one of three fine guidance instruments that is also capable of very precise measurements of the positions of astronomical objects. It often serves as a guider when not being used for astrometry.
- *Repaired:* The Advanced Camera for Surveys (ACS), an optical camera with a wide field of view and imaging capabilities complimentary to those of WFC3; and the Space Telescope Imaging Spectrograph, a versatile ultraviolet–optical instrument capable of both imaging and spectroscopy.

- *Existing:* The Near Infrared Camera and Multi-Object Spectrometer (NICMOS), an instrument with longer wavelength sensitivity than WFC3 and several unique capabilities, such as infrared coronagraphy and polarimetry.

The process of *re*commissioning the new and repaired Hubble instruments actually begins many years before a servicing mission. Scientists at the Space Telescope Science Institute (STScI) and engineers at Goddard Space Flight Center define the astronomical observations and observing sequences so that the recommissioning process can begin immediately after Hubble is repaired. This is a rigorous process in which the verification team examines every detail of the observation or engineering exercise needed to test the instruments. Each procedure undergoes multiple reviews before it becomes a set of instructions ready to execute onboard the telescope. About 150 tests must be passed before the telescope is once again fully ready for general purpose use by the astronomical community.

Release and Checkout

In a typical scenario for each of the servicing missions, at the moment Hubble is released from the shuttle, the process of monitoring the health of the observatory and the status of its new components begins. Progress in the first few weeks is always slow and low key since it takes time for the new equipment to acclimatize to the space environment. Each step is carefully choreographed and has to be verified before moving on. Rushing the process at this point by turning on the high-voltage power supplies could be fatal to Hubble's instruments: the power supplies could short out if there is too much pent-up gas in the instruments remaining from their time on the ground. No matter how well the instruments are "baked out" or preconditioned for flight, there is always some residual gas in the form of hydrocarbons or water that escapes from the instruments when they are first placed into Hubble. It has always been better to be patient and wait until that gas has escaped into the vacuum of space than to risk in haste years of work required to get to this point—no matter the pressure to start producing science!

As the public and astronomical communities eagerly await Hubble's return to service and a glimpse of the first data from the new instruments, the verification team is hard at work. Engineering activities include dumping memory, making sure that information placed in various buffer locations can be read back, exercising mechanisms such as filter wheels, and turning on detectors that operate at low voltages. Once high-voltage operation is safe, calibration lamps and the remaining detectors can be turned on. Detector functionality is a key milestone for the team. If the detectors don't work properly, all of the work involved in getting light from the telescope through the instruments and onto the detectors is for naught. That precious light must be detected and measured! Short exposures with the calibration lamps on and off provide initial indications of detector performance.

Once the team deems the detectors to be functionally fit, it is time to align and focus the instruments so that external light collected by Hubble's primary mirror (and reflected to the instruments by the secondary mirror) can be analyzed. An initial series of tests determines whether the instruments are at their proper focus positions, assesses how well the instrument optics correct for the aberrations introduced by Hubble's primary mirror, and measures the throughput of each instrument. More detailed tests involving images and spectra of astronomical objects soon follow and provide a thorough characterization of the instrument filters, diffraction gratings, and detectors. It is at this point that the true quality of each instrument and an understanding of its scientific potential emerge.[1]

Analyses of these calibration data determine the various tweaks to instrument parameters the Hubble team needs to make to optimize instrument performance. Once the instruments are tuned up, a series of observations called early release observations (EROs) are made to show the public that Hubble is once again producing beautiful images and to provide the astronomical community with examples of the new observations that demonstrate the performance of the refurbished observatory.

For example, the EROs resulting from SM4 were released on 9 September 2009, approximately 14 weeks after the start of recommissioning, as shown in this mosaic of four ERO images (Figure 3.7-1) taken by Hubble's new WFC3.[2]

The verification sequence is highly structured from the standpoint of the ordering of activities, but the exact timing of events is fluid to allow sufficient time for data analysis and modifications to the plan. Not all of the above activities proceed at the same pace for each instrument. Each instrument has its own quirks and special calibrations, and the commissioning of some instrument modes proceeds faster than others. This proved to be fortuitous after SM4. For example, we were fortunate that the commissioning of the visible light channel of the WFC3 had advanced far enough that it was possible to observe Jupiter shortly after an asteroid impacted the giant planet on 19 July 2009, as shown in Figure 3.7-2 taken only four days later. This was the first image from the new camera released to the public, well in advance of the main set of ERO images released in September of that year. Had the impact occurred only a week or two earlier, WFC3 would not have been ready to capture the event.

Figure 3.7-1. After servicing missions and other recoveries from periods of doubt about Hubble's abilities, both NASA and ESA issued new images. Among the first observations made by WFC3 after the final servicing mission, the new images released to the press show objects familiar to Hubble followers ever since the launch in 1990. Clockwise from top left: the Bug Nebula, a dying star, revealing detail not seen in earlier Hubble images; Stephan's Quintet of colliding galaxies; the heart of the giant globular cluster Omega Centauri, long a Hubble favorite; and a star formation region in the vast Carina Nebula that seems to be making a point. (NASA image, credit: NASA, ESA, and Hubble SM4 ERO Team, http://hubblesite.org/newscenter/archive/releases/2009/25.)

Approximately once per year, the STScI issues a call for proposals to astronomers around the world for observing time with Hubble. Nearly 5,100 different people have been official investigators on approved Hubble proposals. There are more than 200 observing programs approved to use its new and repaired

Figure 3.7-2. During Hubble's operational lifetime, two collisions of large bodies with Jupiter have been recorded, heightening awareness that major collisions do take place between planetary bodies in the solar system. Both of these events were many orders of magnitude more powerful than the conflagration resulting from the object that exploded in June 1908 over the Tunguska River Valley in Siberia. This Hubble image, taken 23 July 2009 with the new WFC3, was made four days after the impact and provided high resolution information revealing that the expanding and distorted debris plume was evidently caused not by the collision itself but by turbulence in Jupiter's high atmosphere after the collision. The changing structure of the debris plume, equivalent to a land area larger than the Eastern United States, was similar to patterns seen in the 1994 collision of fragments from comet Shoemaker–Levy 9. Hubble data have also been combined with Earth-based infrared images to assess the vertical structure of the Jovian atmosphere. (NASA image, credit: NASA, ESA, and H. Hammel (Space Science Institute, Boulder, Colorado), and the Jupiter Comet Impact Team, http://hubblesite.org/newscenter/archive/releases/2009/23/image/a/.)

instruments in the current (2010) observing cycle (Figure 3.7-3). These programs address research areas as diverse as the composition of planets in our own solar system to the atmospheres of planets around other stars, from infant galaxies in the early universe to the deaths of nearby stars, and from tenuous wisps of intergalactic gas to exotic objects so dense, like black holes, that not even light can escape. This suite of programs makes use of all of Hubble's instruments that astronauts have worked so hard to fix or install. In many cases, these research investigations make use of more than one instrument, often at the same time.[3]

Messier 83 Imaging

To give some insight into what the refurbished Hubble is capable of observing, I will briefly discuss a few of the early science results obtained soon after SM4. I will start with a beautiful WFC3 image of the galaxy Messier 83

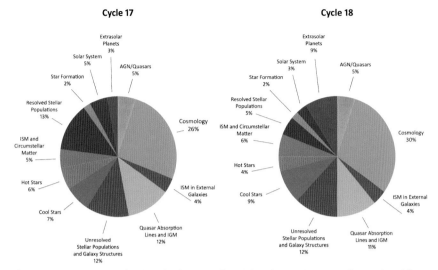

Figure 3.7-3. Diagrams showing the fraction of Hubble's observation time allocated to different science categories for the approved programs in the two most recent observing cycles; abbreviations: ISM, interstellar medium; IGM, intergalactic medium; AGN, active galactic nuclei. (Courtesy of Kenneth Sembach.)

(M83) shown in Figure 3.7-4. Messier 83 is also known as the Southern Pinwheel because of its prominent spiral structure and location in the constellation Hydra. It is about 15 million light years from Earth. Robert O'Connell, a professor at the University of Virginia and a member of the WFC3 Science Oversight Committee, a group of astronomers responsible for providing scientific guidance to the WFC3 instrument development team, produced this figure. The STScI director awarded this group about 150 hours of observing time to be used for imaging galaxies like M83 and demonstrating the science capabilities of the new camera. Some 11 hours of that observing time were used to create the image here.

On the left-hand side of the M83 galaxy image is a ground-based image of the entire galaxy, which is larger than the field of view of the WFC3 outlined in white. Hubble zoomed in on this area of the galaxy, and the result is shown in the multicolor image on right-hand side of Figure 3.7-4. The image demonstrates nicely the power of the camera to reveal the structure of the galaxy in fine detail and in different colors. The instrument channel used here can observe light not only at optical (greenish red) wavelengths but also at ultraviolet (blue) wavelengths. Messier 83 contains numerous young, hot stars that radiate intense ultraviolet light; many of these stars are concentrated in the galaxy's spiral arms. These stars are very young, maybe between only one million and perhaps ten million years old. That span of time passes in the blink of an eye in the history of the cosmos. Hubble eas-

Figure 3.7-4. Two views of the active star birth spiral galaxy M83. The left-hand image is from a special image-concentrating camera on the 2.2 meter telescope of the European Southern Observatory's Max Planck Gesellshaft, observing from the high mountain site above La Silla, Chile. Its 67-million-pixel imager can cover a field as large as the full Moon. The white box indicates Hubble's field of view, imaged in enlarged view on the right by the new WFC3 camera installed during SM4 in May 2009. A key feature of this August 2009 Hubble image is the deep core of the active galaxy, the whitish bar on the far right-hand side. (NASA image, credit: NASA, ESA, R. O'Connell (University of Virginia), B. Whitmore (STScI), M. Dopita (Australian National University), the WFC3 Science Oversight Committee, and the European Southern Observatory, http://hubblesite.org/newcenter/archive/releases/2009/29/.)

ily resolves many of the star clusters in M83; numerous stellar groupings are visible as aggregates of individual stars or as small, tightly knit clusters.

Typical of great spirals, the arms of M83 contain large quantities of glowing hydrogen gas excited by the embedded hot stars. Note the reddish bubbles around many of these stellar objects. There are dozens of bubbles in this particular Hubble image; they are actually spheres of thin gaseous material made visible only by foreshortening. The reddish color of hydrogen emission outlines the remnant bubbles created by supernovae, or stellar explosions, and subsequent cavities carved by strong stellar winds. These energetic processes displace and sweep up the interstellar material, exposing the central stars so that Hubble can examine them. Dark lanes of dust form a patchwork of web-like filaments, obscuring a multitude of other stars and stellar nurseries at these wavelengths.

Images like these provide astronomers with new and very detailed views of how stars interact with their surroundings and of the rate at which new stars form.[4] To be able to see these processes up close in regions where the stars are very young is important because at this stage of stellar evolution the stars are very active. They are altering their surroundings dramatically, carving up the interstellar medium out of which they formed and laying the groundwork for the emergence of new stellar nurseries. By observing the interactions of the stars and gas, Hubble can also age-date the stellar clusters and determine how quickly the surrounding medium is modified.[5]

One of the great strengths of Hubble's new camera is that it has 80 different filters available to study galaxies and other astronomical objects—63 filters at ultraviolet and visible wavelengths and 17 at infrared wavelengths. Astronomers tailor the use of those filters to the type of objects being observed, the wavelengths of light they are interested in studying, and the kinds of information they seek. Five filters were used to produce the image of M83. About half of the programs in the current cycle make use of the camera's extensive filter set.

The Hubble Ultra Deep Field

The power of the WFC3 has been demonstrated in its ability not only to image nearby galaxies but also to detect the most distant galaxies in a patch of the sky known as the Hubble Ultra Deep Field (HUDF), an observing program led by Garth Illingworth at the University of California, Santa Cruz. In Figure 3.7-5 there are a few relatively nearby galaxies but there are also thousands of very distant galaxies that appear as reddish points of light. Essentially every speck of light in this image is a galaxy.

Past observations of this small patch of sky in the southern constellation Fornax, from September 2003 through January 2004 with Hubble's ACS and NICMOS, provided astronomers with their deepest views of the universe, enabling them to look back to a time when the universe was only about 800 million years old, less than 6 percent of its 13.7 billion year age.[6] The new camera (WFC3) improved on this age record by means of its unparalleled ability to detect faint infrared light.[7] Light from the very reddest galaxies in this image began its journey to us only 600 million years after the Big Bang, so there wasn't a lot of time for galaxy formation. The galaxies are very compact and simple from a structural standpoint. They are only a few thousand light-years across compared with the tens of thousands of light-years across for large galaxies like our own in today's universe.

Astronomers are still trying to determine the ages of some of the galaxies in this image (Figure 3.7-5), but because of the galaxies' extreme faintness the analysis is very time consuming. The most distant galaxies are about 10 billion times fainter than could be seen by staring at the sky with our eyes if they were as sensitive to light at these infrared wavelengths as they are to visible light. Until now it was not known whether there would be galaxies more distant than those originally observed with the ACS and NICMOS

Figure 3.7-5. If you see a speck of light in this image with spikes, it is likely a star; if not, it is a galaxy. This image, taken in late August 2009 with WFC3 in near-infrared light, shows that the faintest and reddest objects are indeed galaxies. These galaxies were the deepest seen to that time. The field was the same one exposed by ACS and NICMOS in 2004 but showed the superiority of WFC3, able to image galaxies that formed barely 600 million years after the Big Bang. As with the original Hubble Deep Field exposed in 1995, these fields successfully penetrate farther into deep time and reveal galaxies at younger and younger ages. Hence these images are extremely valuable in assessing how galaxies form. Hubble's operational history is highlighted by successively deeper deep fields. (NASA image, credit: NASA, ESA, G. Illingworth (University of California Observatories [UCO]/Lick Observatory [LO] and University of California, Santa Cruz), R. Bouwens (UCO/LO and Leiden University), and the HUDF09 Team, http://hubblesite. org/newscenter/archive/releases/survey/hubble-ultra-deep-field/2009/31/.)

cameras. It's wonderful to be able to say with confidence that the new WFC3 has shown that more distant galaxies do, in fact, exist in this field.

One might well ask how one could know that light left those galaxies near the beginning of time, and this would be a very good question. The answer is straightforward and a marvelous example of an observation that tells us that the universe is very old indeed.

After 1929, when Edwin Hubble established that there was a relationship between the observed recession velocities of galaxies and their distances, the observed rate of the implied expansion of the universe came to be known as the Hubble Constant. This new conceptual framework directs us to understand that distant galaxies appear redder than they would if they were nearby—their light is redshifted: the larger the distance, the greater the velocity, and thus the greater the wavelength change, or redshift. As the distance increases, the light is shifted redder and redder in color until even the bluest light that could be emitted by the galaxy is red. We can put this to good use in placing limits on the distances for those faint red specks by noting at what wavelengths (or, more properly, in what filters) the galactic light is, or is not, detected. The most distant objects should be the reddest and have little or no light detected at shorter (bluer) wavelengths. This is what is observed by Hubble's new camera for a handful of objects in the Deep Field (Figure 3.7-5). In essence, the older and more distant objects will appear redder than they should because these objects are so old and distant that what we are seeing would otherwise be invisible. The ratio of apparent wavelengths of light to the emitted wavelengths in this case is about a factor of seven or eight, so these are called "redshift seven" or "redshift eight" galaxies.[8]

Finding galaxies like this has been very difficult in the past, even with Hubble. Hubble has another infrared camera (NICMOS), but the new WFC3 is about fifty times more efficient than NICMOS for searching out these high redshift galaxies. It is more sensitive to infrared light and has a larger field of view than NICMOS. We have great expectations for future studies like this with this camera.

Galaxy Cluster Abell 370

Another example of the dividends from SM4 is an improved look at distant clusters of galaxies. In the 1950s, astronomer George Abell used Palomar's 48-inch (1.2 m) Schmidt camera to search for evidence of how galaxies cluster, and he compiled a famous catalogue named now in his memory. Figure 3.7-6 shows a splendid Hubble image of a magnificent cluster located in Cetus the whale, number 370 in Abell's catalog, that was observed with the newly repaired ACS, one of the instruments that astronaut John Grunsfeld fixed by replacing some blown-out electronics boards. In my opinion it is one of the most spectacular images obtained with Hubble and clearly demonstrates how valuable this camera is in revealing nature's secrets.

At the center of this image is a cluster of galaxies located about 4 billion light years away. It consists of a bright yellow central galaxy, surrounded by many smaller elliptical-shaped yellow galaxies. Upon first glance, this image may not look too unusual, but there are some curious arcs of light present, particularly a very striking one on the right-hand side that is extended and looks a little bit like a serpent with a galaxy for a head. Upon further inspection, note that the body of the serpent is actually composed of distorted

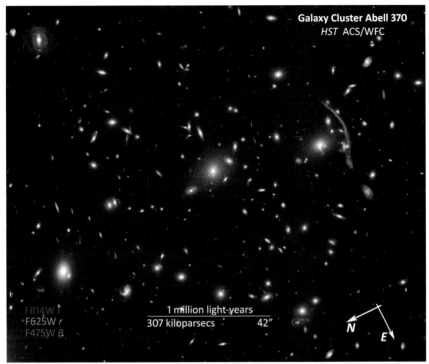

Figure 3.7-6. An Advanced Camera for Surveys (ACS) wide field image taken 16 July 2009 of Abell 370, a cluster of galaxies about 5 billion light years from Earth. The image was constructed from several broad-band exposures. Note the numerous elongated arcs and streaks of light circling the cluster. These are background galaxies that have been gravitationally stretched or lensed by the mass of the cluster. The amount of lensing indicates that there is far more matter in the cluster than is due to the visible galaxies. Astronomers call this missing material dark matter and use the structure of the lensing to determine its amount and distribution. (NASA image; credit: NASA, ESA, Hubble SM4 ERO Team, and Space Telescope-European Coordinating Facility (ST-ECF), http://hubblesite.org/newscenter/archive/releases/2009/25/image/ao/.)

replicas of that galaxy. What's happening here?

The material in Abell 370 is acting as a massive gravitational lens, which bends and distorts the light of background galaxies as the light travels to us. This is analogous to the way an optical lens bends light rays, only instead of the lens being made out of glass or some other common material the lensing is caused by a distortion of the space-time continuum in the vicinity of a massive object, in this case the central cluster of galaxies. Put another way, gravity is bending light as predicted, and now dramatically demonstrated, by Albert Einstein's general theory of relativity.

Lenses not only distort light but also magnify light. So in addition to the light path being changed the light is also amplified, making the background galaxies appear brighter than they would appear had the cluster not been present. For the giant arc in the image, the magnification factor is about 30.[9]

In some cases, this light amplification can be the difference between being able to see an object and having no hope of detecting it. So it is pretty remarkable that the laws of physics give us a chance to see the unseen in special instances such as this.

There are other interesting complexities revealed by this image. When astronomers estimate the amount of mass necessary to bend the light into the arcs and arclets seen, they find that there should be many times more mass than the amount of light present would imply. There must be additional dark matter that is hidden from our view. Even though the exact nature of the dark matter is unknown, the amount and the distribution of the dark matter in a cluster like this can be determined by the positions and the shapes of all of those little arcs and arclets of light. We know from their distribution that there are approximately two mass concentrations in the cluster along this sight line, with a mass of approximately 400 trillion times that of the Sun.[10]

Figure 3.7-7 shows a zoomed-in view of some of the arcs. In the upper right panel you can see very thin blue arcs that require a telescope with superb resolution, like Hubble, to see and resolve. Notice that they are not uniformly bright along their lengths. You can see some of these arcs from the

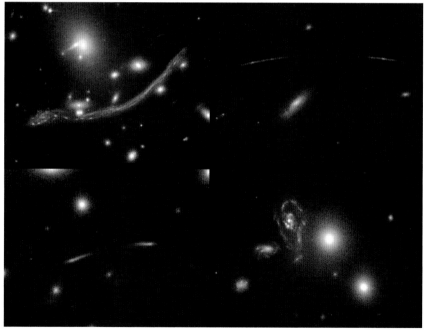

Figure 3.7-7. Details from the ACS Hubble image of galaxy cluster Abell 370, 9 September 2009. (NASA image, credit: NASA, ESA, the Hubble SM4 ERO Team, and ST-ECF, http://hubblesite.org/newscenter/archive/releases/2009/25/image/aq/.)

ground, but you cannot see them in the kind of detail that Hubble's camera reveals. The upper left panel is a close-up view of the spectacular arc mentioned earlier. What we see here is multiple images of that background galaxy superimposed on itself all along this arc. It is actually a picture of that galaxy over and over again as the light gets bent and distorted as it passes through the gravitational lens caused by the cluster. Hubble easily resolves the spiral structure of the lensed galaxy, complete with star-forming regions and spiral arms that may not be too different from those seen in the M83 image, just unimaginably farther away.

Another extremely interesting aspect of this image is the shape of the giant arc along its length. It consists of multiple overlapping images of the same background galaxy. Galaxies in the cluster near the arc distort the shape of the arc; this is gravitational lensing caused by smaller mass concentrations (galaxies) rather than the main mass concentrations in the cluster. Gravitational lensing is working on multiple scales—at the cluster scale to produce the arc and at the galactic scale to distort it.

Hubble Better than Ever

There is no doubt that Hubble has improved with age and is better than it has ever been. The telescope is producing beautiful images and spectra on a daily basis. Approximately two scientific papers based on Hubble data appear in the peer-reviewed scientific literature every day, and thousands of astronomers around the world continue to suggest new and interesting research to conduct every time a new call for proposals is issued. In just the past round of proposals, more than 40 countries and most U.S. states were represented in the proposal pool. There are large observing programs dealing with all of the subjects discussed above—galaxies, the early universe, gravitational lensing—and these are but a few of the many areas in which Hubble continues to make enormous advances in astrophysics. The next essay by David Leckrone provides a nice overview of some of these other areas in which Hubble has had major (and often unanticipated) impact.

Zolt Levay will then relate how, in addition to the science operations work we perform at the Space Telescope Science Institute in support of Hubble, we also have a vibrant education and public outreach program. It has proven very effective, which is gratifying to all of us. At the time of the EROs in September 2009, there was an enormous amount of press coverage in anticipation of how well the refurbished Hubble would work. The same was true at the time of SM4. In those two particular time periods, we estimate that there were more than one billion people worldwide that heard of Hubble or some aspect of Hubble science, through the internet, newspaper articles, or television stories and so on. That's a pretty remarkable number when you think about it. Hubble is an icon and has the ability to inspire people worldwide with its scientific discoveries. I hope it continues to thrive and enhance our lives for years to come.

Notes

1   Since Hubble Servicing Mission 2 in 1997, all of the instruments have had internal optics that correct for the spherical aberration of the telescope's primary mirror. The Corrective Optics Space Telescope Axial Replacement device formerly used to correct for the aberration was removed in SM4 to make room for the new Cosmic Origins Spectrograph to be installed.

2   All of the early release observations released on 9 September 2009 can be found at http://hubblesite.org/newscenter/archive/releases/2009/2009/25/image/ (accessed 15 January 2012).

3   Hubble was launched on 24 April 1990. The 17th cycle of Hubble observations was in progress at the time of this symposium. The 18th cycle was in progress at the time this essay was written.

4   Michael A. Dopita, William P. Blair, Knox S. Long, Max Mutchler, Bradley C. Whitmore, Kip D. Kuntz, Bruce Balick, Howard E. Bond, Daniela Calzetti, Marcella Carollo, Michael Disney, Jay A. Frogel, Robert O'Connell, Donald Hall, Jon A. Holtzman, Randy A. Kimble, John MacKenty, Patrick McCarthy, Francesco Paresce, Abhijit Saha, Joe Silk, Marco Sirianni, John Trauger, Alistair R. Walker, Rogier Windhorst, and Erick Young, "Supernova Remnants and the Interstellar Medium of M83: Imaging and Photometry with the Wide Field Camera 3 on the Hubble Space Telescope," *The Astrophysical Journal* 710 (February 2010):964–978.

5   Bradley C. Whitmore, Rupali Chandar, Hwihyun Kim, Catherine Kaleida, Max Mutchler,Daniela Calzetti, Abhijit Saha, Robert O'Connell, Bruce Balick, Howard E. Bond, Marcella Carollo, Michael J. Disney, Michael A. Dopita, Jay A. Frogel, Donald N. B. Hall, Jon A. Holtzman, Randy A. Kimble, Patrick J. McCarthy, Francesco Paresce, Joseph I. Silk, John T. Trauger, Alistair R. Walker, Rogier A. Windhorst, and Erick T. Young, "Using H2 Morphology and Surface Brightness Fluctuations to Age-Date Stellar Clusters in M83," *The Astrophysical Journal* 729 (March 2011):78–91.

6   National Aeronautics and Space Administration, ESA, S. Beckwith (STScI), and the HUDF Team, "Hubble's Deepest View Ever of the Universe Unveils Earliest Galaxies," STScI press release 2004-07, http://hubblesite.org/newscenter/archive/releases/2004/2004/07/.

7   National Aeronautics and Space Administration, ESA, G. Illingworth (UCO Lick Observatory and the University of California, Santa Cruz), R. Bouwens (UCO Lick Observatory and Leiden University), and the HUDF09 Team, "Hubble's Deepest View Ever of the Universe Unveils Never-Seen-Before Galaxies," STScI press release 2009-31, http://hubblesite.org/newscenter/archive/releases/2009/2009/31/.

8   The record holder for the most distant object detected in the HUDF appears to be a redshift ten galaxy called UDFj-39546284. See R. J. Bouwens, G. D. Illingworth, I. Labbe, P. A. Oesch, M. Trenti, C. M. Carollo, P. G. van Dokkum, M. Franx, M. Stiavelli, V. Gonzales, D. Magee, and L. Bradley, "A Candidate Redshift z ffi 10 Galaxy and Rapid Changes in that Population at an Age of 500 MYr," *Nature* 469(January 2011):504–507.

9   J. Richard, J.-P. Kneib, M. Limousin, A. Edge, and E. Jullo, "Abell 370 Revisited: Refurbished Hubble Imaging of the First Strong Lensing Cluster," *Monthly Notices of the Royal Astronomical Society* 402 (December 2009):L44L48.

10 Richard et al., "Abell 370 Revisited."

8

The Secrets of Hubble's Success

David S. Leckrone

The Hubble Space Telescope (Hubble) is currently, and has been historically, the National Aeronautics and Space Administration's (NASA) most successful and productive scientific mission, as I will quantify later in this essay. There are several primary factors that are characteristic of the observatory and of the mission design overall that led to its remarkable scientific achievements, to its durable popularity, and to its continuing cultural impact.

In his essay, Ed Weiler points out that the NASA Science Missions Directorate is at present (2010) responsible for sixty orbiting missions, but Hubble is the only one that was designed to be serviced by shuttle-based astronauts. The Hubble mission represents a nexus of human and robotic space programs in a way that has greatly benefited both. In my opinion, being bonded to the human space flight program is the root explanation of Hubble's enormous success relative to other robotic missions.

As described by Robert Smith in his wonderful history of the initial development of Hubble, NASA's Office of Space Science began as early as 1969 to seriously consider the relative benefits of launching the Large Space Telescope, as Hubble was then called, on a Titan III rocket or on a manned space shuttle of the kind also being actively pursued by NASA. The Space Shuttle program received approval by the President in 1972. In April of that year John Naugle, head of NASA's Office of Space Science, wrote to George Low, NASA's deputy administrator, that he had concluded that, "it is technically feasible to develop this three-meter optical telescope, and...it can be placed in operation in the decade of the 1980's as an essentially permanent observatory in space through the marriage of automated spacecraft technology and the unique capabilities of the shuttle transportation and maintenance systems."[1] In my view, this realization represented a watershed moment in the history of spaceflight.

I first joined the (re-named) Space Telescope (ST) program in 1976 as the project scientist associated with managing the development of the observatory's scientific instruments. I was deeply excited by the concept that the ST would be the first telescope to be operated in space just like a mountaintop observatory on Earth.[2]

The great mountaintop observatories, like the Palomar Observatory with its 200-inch (-5m) Hale telescope in southern California, have many features that enable continuous scientific excellence and productivity over a long period of time. They have permanent infrastructure, high-quality large-aperture telescopes, and a diversity of interchangeable scientific instrumentation. And of course they are easily accessible to observing astronomers and support staff. The telescope and basic infrastructure is designed to last (with appropriate maintenance and upgrading) for many decades. And over such long periods the scientific power of the observatory naturally advances as the technological capabilities of the scientific instruments evolve. For example, during the twentieth century astronomical imaging capabilities advanced by orders of magnitude as light-sensitive detectors evolved from photographic plates to photography with electronic signal intensification and ultimately to charge-coupled devices (CCDs) and similar two-dimensional solid-state detectors with ever-increasing size, sensitivity, and dynamic range.

The Power of Hubble

Although the Hale telescope has been in active use since 1948, and has now been largely superseded by newer ground-based telescopes with much larger apertures (-8–10 meters), it continued to work at the scientific forefront for many decades because of its ability to be maintained and upgraded and to support ever-more-powerful, interchangeable scientific instruments. The Hubble observatory, married to the Space Shuttle program, was developed specifically as an analog to this paradigm. Get above the clutter and interference of the Earth's atmosphere, put an optically superb telescope in low Earth orbit, base it on an observatory system that is designed to last (with regular maintenance) for decades, not just the 5–10 year norm for robotic missions, develop the capability to regularly upgrade that system technologically, and make it widely available for research by a large community of astronomers and astrophysicists, and what you have is a mountaintop observatory in space, with all the scientific potential that entails.

Another critical factor in Hubble's success is the intrinsic quality of the telescope and supporting spacecraft and ground systems. Although the telescope is only 2.4 meters in aperture, small compared with many mountaintop telescopes, the principle that allows it to detect and measure very faint points of light in the sky is that it should bring the rays of light it collects when pointing steadily at such objects to a very tight focus. The image of a point source should be a very compact bundle the size of which is determined entirely by the size of the telescope's entrance aperture and its associated diffraction pattern—the telescope optics should be "diffraction limited," or very nearly so. When projected on the extremely dark background of the sky observed from orbit, this very tightly focused bundle of light is very efficient in producing a detectable signal above the background noise.

In contrast, the size of a point-source image produced by a mountaintop, large-aperture telescope is dictated primarily by the "seeing" conditions of the intervening atmosphere—the degree to which turbulence and variable refraction in the air smear out the rays of light passing through it. This smeared-out image is projected against a bright sky background produced by the airglow and scattered light intrinsic to the Earth's atmosphere. The light collected even by a 10-meter ground-based telescope is largely wasted and is more difficult to detect above the bright noise background. So, mountaintop telescopes are "seeing limited" rather than diffraction limited. Of course recent advances in adaptive optics have had some success in partially mitigating these atmospheric seeing effects. But adaptive optical systems have their own limitations, and to this day none produce the highly resolved images over wide fields of view to the faint limits of detection at visible wavelengths of which Hubble is capable.

Of course, shortly after its launch in 1990, Hubble's image quality was found to be seriously degraded by a simple grinding error in its primary mirror, resulting in spherical aberration. Only the fact that NASA had created the capability for astronauts to service Hubble in orbit spared the agency and the Hubble program the ignominy of a colossal, embarrassing failure. After Servicing Mission (SM) 1 in 1993, Hubble's "eyesight" became essentially as good as the laws of physics would allow, and its scientific mission could then proceed as planned.

To understand the degree to which this "small" telescope in low Earth orbit with the capabilities described above has advanced the state of observational astronomy, it is interesting to place it in the context of how humankind's capabilities to observe the cosmos with telescopes has advanced over the centuries. One way to do this is to consider how sensitive the telescope and its associated instruments and light detectors are to the photons of light from very faint objects in the sky. To describe quantitatively the apparent brightness of a star relative to other stars or other unresolved point sources of light, astronomers use a system of relative apparent magnitudes:

$$m - m_0 = -2.5\log_{10}(b/b_0),$$

where b is the brightness of the light from the star at a particular wavelength (e.g., visible light) and b_0 is that of some reference standard star. The larger (more positive) a star's magnitude is, the fainter it is (because of the negative sign on the right side of the equation). The smaller the magnitude, the brighter the star is. A commonly used reference standard star is Vega ($\propto$ Lyrae), seen almost directly overhead in the early evening summer sky at mid-northern latitudes. Vega's visible-light apparent magnitude is set at m_0 = 0.0 in this "photometric system."

Astronomers use this logarithmic magnitude system because the range of brightness of planets, comets, stars, external galaxies, quasars, nascent proto-galaxies, and other objects seen out across the cosmos is enormous, spanning many powers of 10. The brightest star we see in the night sky from Earth is Sirius ($\propto$ Canis Majoris). Its visible-light magnitude is m = -1.4. So, Sirius is 3.6 times brighter than Vega. The faintest stars that can be seen with the naked eye in a relatively dark suburban sky under typical conditions is m = 5 or a bit fainter. This corresponds to brightness approximately 0.01 that of Vega.

At the time that Hubble was being developed the largest ground-based telescopes, using photographic detectors, could record stars to a limiting magnitude of about m = 24, or a factor 2.5×10^{10} (the number 25 followed by nine zeros and a decimal point) fainter than Vega. As of today (2010), the faintest objects in the Hubble Ultra Deep Fields observed with Hubble's most modern cameras have a measured magnitude of approximately 30.5. This is a factor 6.3×10^{13} fainter than Vega and a factor of about 6×10^{11} fainter than the faintest stars that can typically be seen with the naked eye from the surface of the Earth. Achieving a limiting magnitude fainter than m = 30 is to astronomy what breaking the sound barrier was to aviation. It is a remarkable achievement, all the more so with a small 2.4-meter telescope.

Galileo Galilei first turned a telescope to the heavens in 1609, and by 1610, with his crude bubble-filled greenish glass lens stopped down to between 0.5- and 1-inch diameter (due to difficulties shaping the lens), he gained an advantage of approximately an order of magnitude in sensitivity to light relative to the capabilities of his naked eye alone. He could resolve faint stars in the Milky Way and could count over three dozen stars in the Pleiades where the casual eye could discern only six or maybe seven. That was the first technological "quantum leap" in the ability of humans to study the cosmos. Over the following four centuries the quality and sensitivity of optical telescopes improved fitfully, gradually, and incrementally. The earliest reflecting telescopes used speculum metal, an alloy of copper and tin, for their primary mirrors. Speculum was a poor reflector of light and was later supplanted by polished glass mirrors coated first with silver and then, by the 1930s, aluminum. Mirrors were first ground to spherical surfaces, but later paraboloid and hyperboloid surfaces did a better job of concentrating light into tightly focused images. The size of telescope apertures, and thus their light-collecting capability, grew. Galileo's small telescopes in the early seventeenth century ranged in aperture from 0.6 to 1.5 inches (~0.02–0.04 m). Later in that century Christiaan Huygens made extensive astronomical use of an 8-inch (~0.2 m) telescope. In the late eighteenth century William Herschel built hundreds of telescopes, the most useful of which was an 18.5-inch (~0.5 m) reflector. In Ireland in 1845 Lord Rosse completed what was by far the largest practical astronomical telescope up to that time: a 72-inch (~1.8 m) reflector dubbed "The Leviathan of Parsonstown" by locals. The first large-aperture mountaintop telescopes in the early twentieth century, ranging up to the famous Hooker 100-inch (~2.5 m) reflector on Mount Wilson near Los Angeles, revolutionized astronomy by making use of photographic plates as light sensors to take deep time exposures of the night sky. By mid-century their successors, such as the 200-inch (~5.1 m) Hale telescope on Mount Palomar, applied electronic sensors and then image tubes. By the end of the century, CCDs greatly expanded the sensitivity range. Today's modern ground-based telescopes have achieved aperture sizes of 400 inches (~10 m). Over the course of four centuries, the sensitivity of ground-based telescopes, due to increased aperture, detector sensitivity, and improved viewing from mountaintop sites, has increased to about 200 million times that of the naked eye. The next major step was to translate these modern telescope and light detection technologies to the space environment, and that was the rationale for Hubble. Hubble undoubtedly has given humankind the greatest single improvement in the capability of a telescope to observe the universe since Galileo. In only two decades it has provided another quantum leap by extending telescope sensitivity in visible light by another factor of almost 100 beyond that of the largest mountaintop telescopes.

Sensitivity to light is only one metric with which to describe the power of a telescope and its instruments. Hubble simultaneously combines in a single observatory unprecedented sensitivity over a wide range of wavelengths (colors), from the far vacuum ultraviolet to the near infrared, while achieving near-diffraction-limited angular resolution over a relatively wide field of view at optical and ultraviolet wavelengths. No mountaintop observatory can compete with Hubble with this combination of virtues.

Even in terms of sensitivity alone, Hubble exceeds the present generation of large-aperture mountaintop telescopes and is likely to be competitive with the enormous 30-meter class telescopes being envisioned for future ground-based observatories. Astronomers at the Space Telescope Science Institute (STScI) have illustrated this point with the calculations tabulated in Figure 3.8-1.[3] These exposure times have been calculated using the Exposure Time

Calculator software programs for specific cameras that are commonly used by observers on these telescopes. One concludes that observations of very faint stars to a signal-to-noise ratio equal to 5 at various wavelengths can be accomplished in mere minutes of Hubble observing time but would require many hours of time with 8-meter mountaintop telescopes. And the stellar brightness (magnitudes) adopted for these calculations don't even approach the $m = 30.5$ limiting magnitude Hubble alone has achieved.

The next major step in telescope and instrument capability beyond Hubble will result from placing an even larger optical telescope in space— the infrared James Webb Space Telescope to be launched in about 2018. A much larger ultraviolet–optical–near-infrared telescope with a 10–15 meter

UVIS Sensitivity: HST Compared with Ground based 8m class telescopes: Exposure Times (seconds) required to reach listed magnitude at S/N = 5
The time estimates for ground observations using average conditions are 4 to 30 times larger than those using UVIS.

	UV	U	B	V	R	I	Z
Vega magnitude	25	25	27	27	27	26.5	25
HST/WFC3/UVIS	3100. (F225W)	1100. (F336W)	2200. (F438W)	1300. (F555W)	2600. (F625W)	3000. (F814W)	1700. (F850LP)
Gemini/GMOS		13500. (u') x12	16200. (g') x7		53100. (r') x20	61200. (i') x20	33300. (z') x19
VLT FORS1		33000. (U$_{Bessel}$) x30	12000. (B$_{Bessel}$) x5	8300. (V$_{Bessel}$) x6	12000. (R$_{Bessel}$) x4	37000. (I$_{Bessel}$) x12	28000. (Gunn Z) x16

Notes: All Calculations done using a Pickles M0V stars, flux normalized in to the filter listed or the closest available one.
WFC3/UVIS: All exposures using CR-SPLIT=2, average background and earthshine.
GMOS: Silver coating, 0.80" seeing, airmass < 1.5. Using 50 percentile sky transparency, and average sky, optimum S/N with aperture ratio = 1.
FORS1: 0.80" seeing, airmass = 1.5, and sky level 10 days from new moon.
All numbers derived using the latest available ETC and ITCs as of March, 25, 2010.

Figure 3.8-1. Comparison of the Hubble Space Telescope's (HST) Wide Field Camera 3 (WFC3) ultraviolet and visible light (UVIS) imager sensitivity to that of other currently operating 8-meter-aperture mountaintop observatories—the Gemini Multi-Object Spectrograph (GMOS) and Very Large Telescope Focal Reducer and Low Dispersion Spectrograph (VLT FORS 1). The exposure time in seconds required to achieve a signal-to-noise (S/N) ratio of 5 on a very faint star or other point source of magnitude "x" at various wavelengths (color filter bandpasses) has been calculated using the observation planning software packages from each observatory. The color filters range from vacuum ultraviolet (UV, only observable from space) through near-ultraviolet (U, observable from the ground), blue (B), visible (V), near-red (R), medium-red (I), to far-red (Z). Other abbreviations: M0V, stellar spectral fluxes normalized to zero magnitude in the V, or visible, bandpass, referring to a catalog of stellar flux standards compiled by A. J. Pickles of the University of Hawaii; CR-SPLIT 2, cosmic ray–split: splitting a camera exposure into two shorter exposures which can be compared to remove the signatures of cosmic rays; ETC, exposure time calculator; ITC, integration time calculator. (Courtesy of Ken Sembach and STScI.)

aperture would also have enormous scientific potential.[4] Hopefully, such a "son or daughter" of Hubble could be launched into a high orbit sometime in the 2020s.

Hubble as a Public Facility

Possibly the most critical factor in Hubble's tremendous success has to do with the manner in which it is operated. Hubble is a public facility observatory, open for use by any astronomer from around the world. Each year an announcement goes out to the worldwide community soliciting research proposals for use of Hubble and its instrumentation. That instrumentation is a complementary set of cameras, spectrographs, and other more specialized devices, such as stellar coronagraphs and interferometers. It is extremely versatile and covers a wide range of performance characteristics, such as sensitivity, resolution, and wavelength coverage. Taken together the Hubble instruments provide essentially a complete toolbox for astronomers to utilize in attacking almost any problem in modern optical astronomy.

Hubble observations have yielded major advances in virtually every area of astronomy and astrophysics. The mechanism by which this is achieved is the open proposal solicitation, peer-review, and selection process that brings into the Hubble research program observers from the entire international astronomical community. The demand for Hubble observing time is huge. In a typical proposal cycle approximately five or six times as much observing time is requested as is available. Many extremely worthy research proposals are rejected each cycle, simply because there is not nearly enough time on the telescope to go around. Only the very best scientific ideas put forward by the community find their way onto Hubble's observing schedule.

The foregoing discussion defines the paradigm from which the Hubble program has achieved its unparalleled scientific success. In summary, NASA (and the American taxpayer) has provided a telescope of superb optical quality and a highly capable set of scientific instruments, placed them in an almost ideal observing site, and regularly maintained them at state-of-the-art performance levels over a long period of time. Management of this complex enterprise has been the responsibility of NASA's Goddard Space Flight Center in Greenbelt, Maryland; and NASA has "thrown open the doors" of this facility to the worldwide community of astronomers, providing access for the very strongest scientific research programs via an open and competitive peer-review process that is managed by the STScI at the Johns Hopkins University in Baltimore. In addition, NASA provides research funding to the successful American competitors to assure they have the means to extract reasonable scientific benefit from their Hubble observations. For those who do not win observing time, the STScI houses an enormous digital archive of all Hubble observational data. These data are usually proprietary for a period of one year to the observers who originally acquired them. After that, they become freely accessible over the internet to anyone who wants to use them. In recent years a substantial fraction of published research papers based on Hubble observations have originated from studies with archival data.

If NASA has provided the world with a long-term observatory in space, the worldwide astronomical community has provided the brainpower and scientific innovation to put Hubble to its very best use. A remarkable consequence is that roughly half of Hubble's most important and impactful scientific achievements have been unexpected. They answer questions Hubble's founders—Lyman Spitzer, John Bahcall, Bob O'Dell, Nancy Roman, and numerous others—did not even know how to ask when they first envisioned the observatory.

Expected and Unexpected Science

Of course Hubble was originally "sold" to the scientific community and to the federal government on the basis of a suite of important scientific problems that could be addressed from only the vantage point of a large-aperture telescope above the atmosphere. In the late 1960s and early 1970s groups of

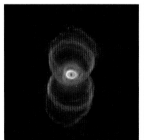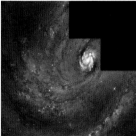

Figure 3.8-2. The Hourglass Nebula, a planetary nebula formed in the final stages of a star's evolution. (NASA image; credit: R. Sahai, J. Trauger (JPL), WFPC2 Science Team, NASA; http://hubblesite.org/newscenter/archive/releases/1996/07/.)

Figure 3.8-3. Spiral galaxy M100 in the Virgo cluster of galaxies. Measuring the distance to the Virgo cluster using Cepheid variable stars observed in this and other cluster galaxies was the cornerstone of the Hubble Key Project to calibrate the distance scale and age of the universe. (NASA image; credit: NASA, J. Trauger (JPL), WFPC2 Science Team; http://hubblesite.org/newscenter/archive/releases/galaxy/spiral/1994/49/image/c/.)

Figure 3.8-4. Center of the giant galaxy M87, highlighting relativistic jet shooting out from its energetic core. The core contains a supermassive black hole approximately three billion times the mass of the sun. (NASA image; credit: NASA and the Hubble Heritage Team (STScI/AURA); http://hubblesite.org/newscenter/archive/releases/galaxy/quasar_active-nucleus/2000/20/.)

astronomers conducted and publicized detailed studies of what such a space telescope could be expected to accomplish scientifically. Perhaps the most influential of these was the final report of a National Academy of Sciences study chaired by Lyman Spitzer, affectionately and humorously dubbed "Chairman Spitzer's Little Black Book."[5] Essentially all of the exemplar scientific objectives of Hubble originally put forward to justify its scientific importance have now been achieved. Examples of discoveries and scientific achievements that were anticipated for Hubble prior to its launch can be seen in Figures 3.8-2, 3.8-3, and 3.8-4.

On the other hand the biggest excitement about Hubble science, the outcome of an observatory design philosophy based upon "conscious expectation of the unexpected," derives from the surprises Mother Nature has revealed in Hubble observations. Examples of discoveries and scientific achievements of Hubble that were completely unexpected prior to its launch can be seen in Figures 3.8-5, 3.8-6, and 3.8-7.

An original Hubble program objective was to measure the rate of expansion of the universe at the present time, the Hubble Constant H_0, to an accuracy of about 10 percent. This value, in turn, could be used in conjunction with cosmological models to derive the expansion age of the universe. The Hubble "H_0 Key Project" succeeded admirably in achieving this objective, $H_0 = 72 \pm 8$ kilometers second^{-1} megaparsec^{-1}.[6] However, the resulting

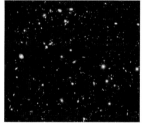

Figure 3.8-5. Dark sooty remnants in the upper atmosphere of Jupiter left by multiple impacts of fragments of Comet Shoemaker-Levy 9 in 1994. These Hubble observations heightened public awareness of the potential for future asteroid or comet impacts on Earth. (NASA image; credit: NASA, Heidi Hammel (Space Science Institute), and HST Comet Team; http://hubblesite.org/newscenter/archive/releases/solar-system/jupiter/1994/1994/34/.)

Figure 3.8-6. The Bullet Cluster, two colliding galaxy clusters in which ordinary matter and dark matter have been segregated, as imaged from a ground-based optical telescope as well as from Hubble and from Chandra X-ray Observatory (CXC). (NASA image; credits: X-ray: NASA/CXC/CfA/M. Markevitch et al.; optical: NASA/STScI, Magellan/U. Arizona/D. Clowe et al.; lensing map: NASA/STScI; ESO WFI; Magellan/U. Arizona/D. Clowe et al.; http://chandra.harvard.edu/photo/2006/1e0657/.)

Figure 3.8-7. The Hubble Extreme Deep Field (XDF) made by stacking many Hubble images of the same field in the sky acquired in the period 2003–2011 with the ACS and WFC3 cameras. This is humankind's deepest look across the universe and back in time to date. (NASA image; credit: NASA, ESA, G. Illingworth, D. Magee, and P. Oesch (University of California, Santa Cruz), R. Bouwens (Leiden University), and the HUDF09 Team; http://hubblesite.org/newscenter/archive/releases/cosmology/2012/2012/37/.)

age calculations were strangely young—somewhere between 9 and 12 billion years—whereas the oldest stars in the universe had ages calculated from the theory of stellar interiors and evolution closer to 15 billion years or older. The universe appeared to be younger than the oldest stars within it, a paradox cosmologists had faced before.

This time, Mother Nature resolved the paradox. In perhaps the greatest achievement in astronomy in the past two decades, Hubble observer Adam Riess and his team, collaborating with scientists at ground-based telescopes, found that the universe is not only expanding at the present time but has been accelerating in its expansion over time.[7] This result was verified independently by a large team of astronomers, headed by Saul Perlmutter, using both Hubble and ground-based telescopes. This finding was completely contrary to the expectation, based on our current understanding of ordinary gravity—that the expansion of the universe should steadily slow down under the mutual gravitational tug of all the matter within it. The driving force or energy source causing the universal expansion to speed up is, today, a complete mystery. But it has been given a name—"dark energy"—and represents over 70 percent of the total mass–energy content of the universe. If we take into account the effects of dark energy and the fact that the universe has been accelerating rather than slowing down for the past 4–5 billion years, we derive an expansion age for the universe in the range 13–14 billion years. The theory of stellar evolution has also now been refined, yielding ages for the oldest stars in that same range. Finally, observations of the cosmic microwave background radiation, particularly with the Wilkinson Microwave Anisotropy Probe (WMAP) space mission, yield an even more precise age measurement (starting with the Big Bang) for the universe of 13.7 ± 0.2 billion years.[8] This general agreement of age estimates and other cosmological parameters obtained independently from multiple observatories using diverse techniques has led to a "concordance" cosmological model for the universe that is now widely accepted by astronomers—the "λCDM," or cold-dark-matter-with-a-cosmological-constant model.

The foregoing discussion illustrates the power of Hubble as a public observatory facility in space enhanced with ever advancing state-of-the-art technology. Hubble observations first raised a perplexing problem and then, later, took the lead in finding the solution to the problem in a completely unexpected discovery.

Advancing Science Step-By-Step

Hubble's ability to advance scientific understanding in multiple steps, as the performance capabilities of its onboard scientific instruments are advanced through servicing, is beautifully illustrated via its discoveries of disks and rings of gas and dust surrounding many stars beyond the Sun. These are the environs in which systems of planets around other stars (exoplanets) must be forming or have already formed.

That stars are sometimes surrounded by dusty structures was demonstrated in observations taken in 1983 by the Infrared Astronomy Satellite (IRAS).[9] The dust surrounding a star is heated by absorbing starlight. It re-emits that energy as thermal, infrared radiation. These early space-based observations revealed that some stars are excessively bright at infrared wavelengths, relative to other stars of similar type, due to thermal radiation from the dusty structures surrounding them. In 1984 astronomers Brad Smith and Richard Terrile used a "coronagraph" (which I will describe shortly) attached to a ground-based telescope to observe visually one of these dusty structures—a large, flattened disk of dust particles surrounding the nearby southern hemisphere star β Pic (beta Pictoris).[10] Although the resolution of these observations was low, the size and shape of the disk, and the inferences that could be drawn about the properties of the dust it contained, allowed the observers to surmise that it was a place where planet formation may have recently taken place. Over the next few years, Smith and Terrile examined over a hundred other stars with their coronagraph but were unable to see any other examples of disks or other dusty structures.

As with so many other subjects in observational astronomy, the detailed visual study of proto-planetary disks and other dusty structures around stars, so challenging for ground-based telescopes, proved relatively straightforward for Hubble. Hubble's first "workhorse" camera, the optically corrected Wide Field and Planetary Camera 2 (WFPC2), installed during the first servicing mission in 1993, almost immediately provided direct visual pictures of dusty structures surrounding many nearby stars. Astronomer C. R. O'Dell, using the WFPC2 to observe the Orion Nebula for entirely different purposes, stumbled upon dozens of stars within the nebular complex that were clearly seen to be surrounded by disks of gas and dust (Figures 3.8-8 and 3.8-9).[11] O'Dell called them proplyds, an abbreviation of proto-planetary disks. These undoubtedly resemble what our own solar system must have looked like in the early phases of its formation. Today, our own Sun is still surrounded by a flattened disk of particles ranging in size from fine dust to objects the size of Pluto or larger in a region beyond Neptune that we call the Kuiper Belt. It is within such a disk that our system of planets formed, and it was reasonable to conclude based on Hubble observations that planetary systems are also evolving within the disks that are now known to be relatively common around other stars. But could this idea somehow be verified by direct observations of the shape, size, and structural details of these disks or perhaps even by direct visual observation of the planets themselves?

Taking a picture of an extremely faint planet in the intense glare of the star around which it orbits is one of the most challenging observations astronomers can attempt. It has been likened to trying to take a picture of a firefly next to the beam of a lighthouse. To reduce the glare of starlight so that a long-time exposure of the faint surrounding environment can be acquired, astronomers use a coronagraph. In this kind of camera an opaque spot or blade is placed precisely over the image of the star focused on the light detector within the camera so that its glare does not swamp the photons of light being detected from the surrounding circumstellar region.

During the second servicing call to Hubble in 1997, two new scientific instruments were installed—the Space Telescope Imaging Spectrograph (STIS) and the Near Infrared Camera and Multi-Object Spectrometer (NICMOS). Both of these instruments contained relatively simple coronagraphs. For the first time astronomers using STIS and NICMOS could spatially resolve underlying structure within circumstellar disks. And the results were revelatory.

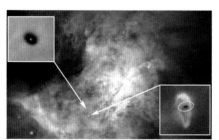

Figure 3.8-8. The Orion Nebula complex observed by astronomer C. R. O'Dell with Hubble's WFPC2 in 1994. Two examples of proplyds (proto-planetary disks) are indicated. The inset at lower right shows a circumstellar disk (circled) surrounded by a bubble of gas being ejected from the proplyd as it is heated by radiation from a nearby hot massive star. The proplyd at upper left is quiescent and appears to be shadowed and protected from intense stellar radiation; it is the more likely of the two to retain the raw material from which planets might form. (NASA images; credits: main image, NASA, C. R. O'Dell and S. K. Wong (Rice University); inset upper left, Mark McCaughrean (Max-Planck-Institute for Astronomy), C. Robert O'Dell (Rice University), NASA; http://hubblesite.org/newscenter/archive/releases/nebula/emission/1995/1995/45/; inset lower right, C. R. O'Dell (Rice University), NASA; http://hubblesite.org/gallery/album/nebula/pr1994024b/.)

Figure 3.8-9. A beautiful example of a circumstellar disk in the Orion Nebula complex seen edge-on. The disk is actually a torus (shaped like a donut), with its inner region cleared of dust. The central star cannot be seen directly, but its light is visible as a reddish glow as it is scattered from dust particles above and below the disk. (NASA image; credit: J. Bally (University of Colorado) and H. Throop (SWRI); http://hubblesite.org/newscenter/archive/releases/nebula/emission/2001/2001/13/image/b/.)

Figure 3.8-10 shows two examples of coronagraphic observations of dusty, circumstellar structures made with both instruments. In the two images on the left, we find circling the star HD 141569 a flattened disk containing multiple gaps. One is reminded of the gaps in Saturn's rings that are caused by the localized gravitational influence of small, "shepherding" satellites. By analogy scientists surmise that the gaps in the disk of HD 141569 may result from the gravitational influence of proto-planets on the surrounding dust particles. The proto-planets sweep the disk clean of smaller particles within or near

Figure 3.8-10. Coronagraphic observations of circumstellar dusty structures around two stars (HR 4796A, left column; HD 141569, right column) as obtained with the Near-Infrared Camera and Multi-Object Spectrometer (NICMOS; top row) and the Space Telescope Imaging Spectrograph (STIS; bottom row) instruments mounted on Hubble in 1997. For the first time structural details—the internal architecture—of these systems could be seen. Gaps in dusty disks and ring structures are not stable unless they are maintained by gravitational interactions with planets or proto-planets that cannot be seen directly in these images. (NASA images; credits: NICMOS, top right: Alycia Weinberger, Eric Becklin (UCLA), Glenn Schneider (University of Arizona), NASA; top left: Brad Smith (University of Hawaii), Glenn Schneider (University of Arizona), NASA; http://hubblesite.org/newscenter/archive/releases/star/protoplanetary-disk/1999/03/; STIS, bottom left and right: C. Grady, B. Woodgate (Goddard Space Flight Center), STIS Science Team.)

their orbits. In the absence of such shepherding proto-planets, the gaps are not stable structures and could not continue to exist for very long.

The star shown in the two right-side images of Figure 3.8-10, HR 4796A, is clearly surrounded by a ring. Again, such a structure is not dynamically stable and could not exist on its own without the gravitational influence of one or more nearby proto-planets. In both of these cases, we do not directly observe the planets themselves but infer their existence as being necessary to explain structures that could not exist for very long without their gravitational influence.

A more sophisticated coronagraphic camera was inserted in Hubble during SM 3B in 2002, as a subsystem of a new, high-tech instrument called the Advanced Camera for Surveys (ACS). This coronagraph was more effective than its predecessors in blocking out the glare of a central star, and using multiple color filters it could obtain coronagraphic images at multiple wavelengths. The latter capability was especially important in discerning the physical properties of the particles of dust that constitute circumstellar dusty structures. In one of the more heroic achievements of the final Hubble servicing mission, SM4, in 2009, the ACS was restored to operation after having suffered multiple electronic failures. Unfortunately, its High Resolution Channel (HRC) in which the coronagraph is housed could not be brought back to operational life. Prior to the loss of the ACS coronagraph, however, it

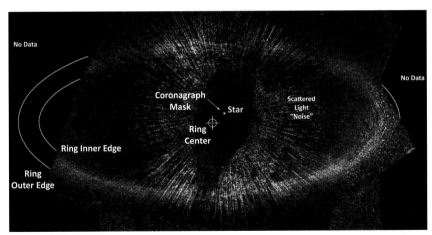

Figure 3.8-11. Coronagraphic observation in 2004 of the circumstellar dust ring surrounding the star Fomalhaut, obtained with the Advanced Camera for Surveys/High Resolution Channel installed on Hubble in 2002. The geometrical center of the ring is offset from the position of the star (obscured by a coronagraphic mask), suggesting the ring may be maintained by gravitational interaction with a planet in an eccentric elliptical orbit. (NASA image; credit: NASA, ESA, P. Kalas, J. Graham (University of California, Berkeley) and M. Clampin (Goddard Space Flight Center); http://hubblesite.org/newscenter/archive/releases/star/protoplanetary-disk/2005/2005/10/.)

achieved another major step forward in the study of circumstellar structures and the search for extra-solar planets, namely the first direct visual image of a planet in orbit around another star.

In 2004 astronomer Paul Kalas and his collaborators used the ACS–HRC coronagraph to produce a remarkable image of a dusty ring surrounding the very bright, nearby, southern hemisphere star Fomalhaut (∝ Piscis Austrini).[12] It is believed that the ring is composed of the debris left over from a prior period of planet formation. The location of Fomalhaut itself is marked in the diagram (Figure 3.8-11), but in this picture the coronagraphic disc blocks the image of the star. The geometrical center of the ring is offset from the star's position. If the ring is produced by the gravitational influence of a planet, the astronomers infer from this that the planet might be in an eccentric elliptical orbit.

In 2006 Kalas and his collaborators observed Fomalhaut for a second time (Figure 3.8-12). While carefully comparing the 2004 and 2006 observations, they achieved yet another amazing Hubble discovery: a direct image of a planet, dubbed Fomalhaut b, moving in its orbit around the star.

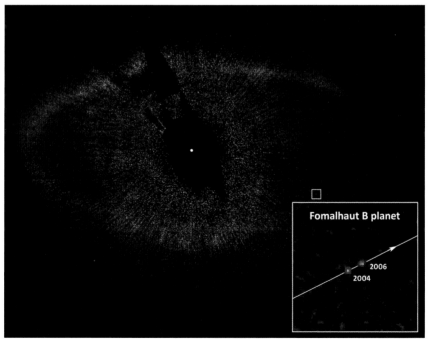

Figure 3.8-12. Similar to the earlier image of Fomalhaut, this 2006 observation has been combined with the 2004 data. Comparison of the observations separated in time by two years reveals the orbital motion of a planet that has been named Fomalhaut b. (NASA image; credit: NASA, ESA, P. Kalas (University of California, Berkeley), et al.; http://hubblesite.org/newscenter/archive/releases/star/protoplanetary-disk/2008/39/.)

Thus, beginning with the discovery of proplyds in 1994 and culminating in the first visual images of an extra-solar planet in 2004–2006, the Hubble observatory opened and greatly advanced an entirely new field of observational research—the study of the detailed sizes, shapes, compositions, and internal structures of the environments around stars in which planets are forming or have been formed. This was possible over a period as short as a decade only because of the nature of Hubble itself as a long-lived, serviceable, public space observatory.

Hubble's Metrics

Each year for the past 19 years, Greg Davidson (former program manager for HST Instrument Development at NASA Headquarters) has compiled the *Science News* metrics that are used to measure NASA's contributions to worldwide scientific discovery and technological achievement.[13] The current release of Davidson's analysis (Figure 3.8-13) spans a 38-year period from 1973 to 2010 and covers all fields of science. He bases his compilation on the annual end-of-year review of the most important discoveries in all fields of science as assessed by the editors of *Science News* magazine. The analysis assigns numerical points to a NASA mission or program based on the degree to which it contributed to each "important scientific finding." For example, for 2010, NASA's missions and programs were responsible for 8.4 percent of worldwide scientific discover-

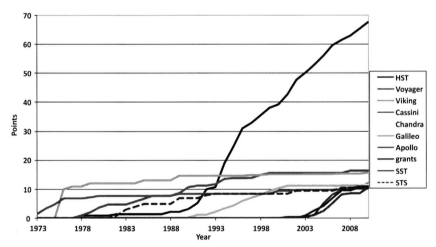

Figure 3.8-13. Cumulative point scores over a period of four decades in the Davidson *Science News* metric for various NASA science missions and programs, indicating the relative contributions made to the "most important science discoveries of the year" in all areas of science, worldwide, as judged by the editors of *Science News* magazine. Abbreviations: HST, Hubble Space Telescope; SST, Spitzer Space Telescope; STS, Space Transportation System. (Image courtesy of Greg Davidson, former program manager for HST Instrument Development, NASA headquarters.)

ies. Of these, 7.0 percent came from Space Science and 1.4 percent from Earth Science. Leading the field was NASA Astrobiology, which produced 1.7 percent of discoveries; in second place was Hubble, producing 1.3 percent.

This is very typical of Hubble performance year after year in the *Science News* metrics. Hubble has been at or near the top ranking of NASA missions and programs ever since its faulty optics were corrected in 1993. As demonstrated here, many NASA missions have performed well in the *Science News* metrics over periods of a few years, typically during their early years of prime operations. But none have sustained scientific excellence over nearly two decades in the way Hubble has. In this respect Hubble has been NASA's most valuable and productive science mission and it remains so today.

Over the four-year period 2007–2010 an average of 697 scientific papers per year based on Hubble observations have been published in refereed professional journals. In 2010 the number of refereed Hubble publications was 704, second only to the 724 published in 2007. This exceeds the annual publication rate of any other comparable astronomical observatory either in space or on the ground.

Over 5,000 astronomers from around the world have obtained observational data with the Hubble, and over 10,000 astronomers have used data from the Hubble archives. The demand for Hubble observing time far exceeds the amount of time available. Typically the oversubscription rate is from 5:1 to 6:1. However, the demand always grows in the first year or two after new instruments are inserted or old instruments are repaired on a Hubble servicing mission. The post-SM4 era is no exception. In the most recent observing proposal submission cycle (cycle 18) astronomers submitted 1,051 proposals, the second highest number ever, and the oversubscription rate of 11:1 was a new record. After 20 years of operation Hubble shows no signs of abating in terms of its importance for research to the scientific community, its scientific productivity, or the excellence of its research programs—extending dramatically the frontiers of knowledge on a regular basis.

Hubble's Bequest

At the present time no serviceable observatories in space are included in NASA's long-term plans for the future. This is ironic considering the unprecedented record of scientific achievement of the Hubble. It can also be argued that NASA as an agency has benefited immensely from Hubble's record of unparalleled public support. Hubble is the "crown jewel" of the agency, not just of Space Science. One can only hope that the paradigm of serviceable, broadly capable, public facility observatories in space will not die with the inevitable demise of Hubble.

Although the Space Shuttle Program has ended, it would be possible in principle to place an observatory similar to Hubble in an orbit at the same high inclination, 51.6°, as that of the International Space Station (ISS), though perhaps at a higher altitude than that of the ISS so that it could be serviced by ISS-based astronauts and cosmonauts. Hubble cannot be moved to such a location from its current 28.5° orbit because of the enormous amount of rocket fuel that would be required.

The German rocket scientist, Hermann Oberth, and his colleague, Hermann Noordung, in describing the potential practical and scientific advantages of a space station in the 1920s envisioned that such human-tended orbiting platforms would naturally serve as a base for the construction and operation of large reflecting astronomical telescopes.[14] To date the ISS's potential in this area remains unexploited.

There are sound technical reasons (e.g., the need for very cold, thermally stable space environments and avoidance of the large angular obstruction of the sky by the Earth's disk) for placing major astronomical facilities in high orbits far away from Earth. Hence, the gravitationally semi-stable L2 Lagrange point one million miles from Earth in the direction away from the Sun has proven a popular location for highly sensitive astronomical facilities, such as WMAP and the James Webb Space Telescope. Perhaps as human space travelers, or their robot surrogates, are able to venture beyond low Earth orbit, a next-generation Hubble will rise again in the form of an even more powerful, very large aperture, serviceable ultraviolet–optical–near infrared "mountaintop" observatory in space.

Notes

1   Robert W. Smith, *The Space Telescope: A Study of NASA, Science, Technology, and Politics*, with contributions by Paul A. Hanle, Robert H. Kargon, and Joseph N. Tatarewicz (New York: Cambridge University Press, 1989):85.

2   The first mission to provide direct real-time observing for guest observers on a formal basis was the International Ultraviolet Explorer.

3   Kenneth Sembach, M. Mountain, and the Hubble Team, Space Telescope Science Institute, 2009, personal communication.

4   M. Postman, T. Brown, K. Sembach, M. Giavalisco, W. Traub, K. Stapelfeldt, D. Calzetti, W. Oegerle, M. Rich, P. Stahl, J. Tumlinson, M. Mountain, R. Soummer, and T. Hyde, "Science Drivers and Requirements for an Advanced Technology Large Aperture Space Telescope (ATLAST): Implications for Technology Development and Synergies with Other Future Facilities," in *Space Telescopes and Instrumentation 2010: Optical, Infrared, and Millimeter Wave*, ed. J. M. Oschmann Jr., M. C. Clampin, and H. A. MacEwen, Proceedings of SPIE (Society of Photographic Instrumentation Engineers) 7731 (2010):77312K–77312K-12.

5   Lyman Spitzer, Chairman, Scientific Uses of the Large Space Telescope, "Report of the Space Science Board ad hoc Committee on the Large Space Telescope" (Washington, D.C.: National Academy of Sciences, 1969).

6   W. Freedman, B. Madore, B. Gibson, L. Ferrarese, D. Kelson, S. Sakai, J. Mould, R. Kennicutt Jr., H. Ford, J. Graham, J. Huchra, S. Hughes, G. Illingworth, L. Macri, and P. Stetson, "Final Results from the Hubble Space Telescope Key Project to Measure the Hubble Constant," *Astrophysical Journal* 553 (May 2001):47–72.

7   A. Riess, A. Filipenko, P. Challis, A. Clocchiatti, A. Diercks, P. Garnavich, R. Gilliland, C. Hogan, S. Jha, R. Kirshner, B. Leibundgut, M. Phillips, D. Reiss, B. Schmidt, R. Schommer, C. Smith, J. Spyromilio, C. Stubbs, N. Suntzeff, and J. Tonry, "Observational Evidence from Supernovae for an Accelerating Universe and a Cosmological Constant," *Astronomical Journal* 116 (September 1998):1009–1038; S. Perlmutter, G. Aldering, G. Goldhaber, R. A. Knop, P. Nugent, P. G. Castro, S. Deustua, S. Fabbro, A. Goobar, D. E. Groom, I. M. Hook, A. G. Kim, M. Y. Kim, J. C. Lee, N. J. Nunes, R. Pain, C. R. Pennypacker, R. Quimby, C. Lidman, R. S. Ellis, M. Irwin, R. G. McMahon, P. Ruiz-Lapuente, N. Walton, B. Schaefer, B. J. Boyle, A. V. Filippenko, T. Matheson, A. S. Fruchter, N. Panagia, H. J. M. Newberg, and W. J. Couch, "Measurements of Omega and Lambda from 42 High-Redshift Supernovae," *Astrophysical Journal* 517 (June 1999):565–586.

8   N. Jarosik, C. Bennett, J. Dunkley, B. Gold, M. Greason, M. Halpern, R. Hill, G. Hinshaw, A. Kogut, E. Komatsu, D. Larson, M. Limon, S. Meyer, M. Nolta, N. Odegard, L. Page, K. Smith, D. Spergel, G. Tucker, J. Weiland, E. Wollack, and E. Wright, "Seven-Year Wilkinson Microwave Anisotropy Probe (WMAP) Observations: Sky Maps, Systematic Errors, and Basic Results," *Astrophysical Journal Supplement Series* 192 (February 2011):1–15.

9   D. Backman, F. Gillett, and F. Low, "IRAS Observations of Nearby Main Sequence Stars and Modeling of Excess Infrared Emission," *Advances in Space Research* 6 (1986):43–46.

10 B. Smith and R. Terrile, "A Circumstellar Disk around β Pictoris," *Science* 226 (21 December 1984):1421–1424; B. Smith and R. Terrile, "The Beta Pictoris Disk: Recent Optical Observations," *Bulletin of the American Astronomical Society* 19 (June 1987):829.

11 C. O'Dell and K. Wong, "Hubble Space Telescope Mapping of the Orion Nebula. I. A Survey of Stars and Compact Objects," *Astronomical Journal* 111 (February 1996):846–855; M. McCaughrean and C. O'Dell, "Direct Imaging of Circumstellar Disks in the Orion Nebula," *Astronomical Journal* 111 (May 1996):1977–1986.

12 P. Kalas, J. Graham, and M. Clampin, "A Planetary System as the Origin of Structure in Fomalhaut's Dust Belt," *Nature* 435 (23 June 2005):1067–1070; P. Kalas, J. Graham, E. Chiang, M. Fitzgerald, M. Clampin, E. Kite, K. Stapelfeldt, C. Marois, and J. Krist, "Optical Images of an Exosolar Planet 25 Light-Years from Earth," *Science* 322 (November 2008):1345–1348.

13 C. Christian and G. Davidson, "The Science News Metrics," *Organizations and Strategies in Astronomy* Volume 6, *Astrophysics and Space Science Library* 335:145–156, http://www.springerlink.com/content/978-1-4020-4055-9/#section=428434&page=1&locus=0.

14 The original references to Oberth's and Noordung's seminal publications are Hermann Oberth, *Wege zur Raumschiffahrt* [Ways to Spaceflight] (Munich and Berlin: R. Oldenbourg, 1924) and Hermann Noordung, *Das Problem der Befahrung des Weltraums—der Raketen-Motor* (Berlin: R. Schmidt, 1928). Noordung's book has been translated into English in the NASA History series: Hermann Noordung, *The Problem of Space Travel: The Rocket Motor* (Washington, D.C.: NASA SP-4026, 1995). It is also on line at http://history.nasa.gov/SP-4026/cover.html. The work of the German school of space stations, including Oberth's and Noorung's ideas, is also discussed in Roger D. Launius, *Space Stations: Base Camps to the Stars* (Washington, D.C.: Smithsonian Books, 2003).

9

Creating Hubble's Imagery

Zoltan Levay

Much of the discussion by the authors of this volume has concentrated on the impact of the Hubble Space Telescope (Hubble) scientifically and technologically, which has been immense. Here I will concentrate specifically on the imagery because that is what I am most familiar with. I think one reason Hubble has become such an icon of space technology and astronomy, and so well received and liked, is the steady stream of compelling images that we have been able to produce. The images are reproduced frequently and widely around the world and used in news, magazines, documentaries, popular culture, and online.

These images are not the results of a snapshot camera. Hubble is a complex device, and the images are a fortunate by-product of its science observations. Hubble was, of course, designed to obtain data for cutting-edge science by means of a marvelous suite of instruments in an observatory on the leading edge of science and technology. The familiar color images that the public enjoys, however, are a result of being able to take these scientific data and produce visually compelling images by making a few different choices.

A conscious effort is required to put these images together, including the subjective choices necessary as part of the production procedures. Here I will explore those subjective choices and their relevance, emphasizing that the images are rich and compelling not because of how we manipulate them but because of the quality of the underlying data. Figure 3.9-1 shows the raw data, essentially what we start with when we begin to create Hubble Heritage images.

Imaging Tools and Techniques

The science instruments are digital cameras, not too different from the technologies you find in a consumer level digital camera. But unlike the color detectors in commercial cameras, the WFPC2 utilizes monochromatic detectors yielding images in black and white. There is no color information, per se, in the images. The only color information we have is the filter that was used in front of the detector. These filters isolate very specific and carefully designed

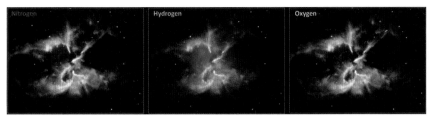

Figure 3.9-1. Three images of the NGC 2818 Planetary Nebula as they came from the instrument data flow: black and white digital images, known also as grayscale. The images were observed with Wide Field Planetary Camera 2 (WFPC2) not long before it was removed during the last servicing mission. (NASA image; credit: NASA, ESA, and the Hubble Heritage Team (Association of Universities for Research in Astronomy/Space Telescope Science Institute [AURA/STScI]); http://heritage.stsci.edu/2009/05/.)

ranges of colors from the electromagnetic spectrum that the detector will see.

As Ken Sembach noted in his essay, there is a large selection of filters, designed for specific science goals, inside the cameras. They provide numerous ways of sampling the light to be able to study the things that astronomers want to study. The filters chosen in this example isolate and sample very specific colors of light that are emitted by particular chemical elements in certain physical conditions. In this example, the filters isolate the light of hydrogen, nitrogen, and oxygen. It may not be immediately obvious that there is anything different in these images, but if examined carefully the differences become clearer.

We can assign color to these separate images and combine them digitally with image processing software. Figure 3.9-2 shows what it looks like when we assign the colors that the filters transmit. The light of nitrogen is red—actually very close to the same color as the hydrogen light. The oxygen light is blue-green or cyan. If we then combine those images, we come up with an image like the one at lower right of Figure 3.9-2, and we see different features in the object pop into better view. We see areas where the nebula is shining mostly in red light, and we see areas with the nebula shining in blue light, and they are distinct. We can learn some things from this image, and it is fairly attractive. These color differences result from different physical processes occurring within the nebula.

There is another approach, however, which is to use the full color capability that we have available. Every color technology uses a three-color model to reproduce color images. For the images made for TV screens or computer screens, the three colors are red, green, and blue—known as the additive primaries. The brightness value of each color component combines in ratios with the other colors to produce the vast range of hues visible to human eyes. In our first go at this image (Figure 3.9-2), we used just two colors: red and blue. If we assign those colors a little bit differently—say that the hydrogen will be shown

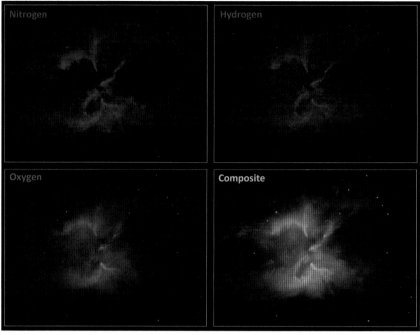

Figure 3.9-2. The image from Figure 3.9-1 has been modified to reflect the visible colors transmitted by the filters. The hydrogen light that we are sampling is red light. (NASA image; credit: NASA, ESA, and the Hubble Heritage Team (AURA/STScI); http://heritage.stsci.edu/2009/05/.)

in green, the nitrogen will be shown in red, and the oxygen will be shown in blue—then we end up with quite a different image (Figure 3.9-3). I think it is a visually more attractive image. It actually shows more distinct colors and shows in greater detail the features within this object at which we are looking.

In fact I would say that this image (Figure 3.9-3) is a more faithful representation of the data than the first example (Figure 3.9-2), even though that original red–blue version is closer to what our eyes would see if we could look directly through Hubble. Human eyes did not evolve to be very good at seeing this sort of thing. If we use instruments specifically optimized for studying these phenomena, then the best representation of the data is one that renders visible, as much as possible, the information inherent in those data. In the first example we masked a lot of structure by rendering the hydrogen and nitrogen light in the same color. When we separate the colors and make use of the full capability of the technology, we are able to show that there are different physical conditions in different places in the nebula. It is actually more informative in that it renders visible more information about the physical processes occurring in the nebula.

The repeated image in Figure 3.9-4 was observed with the newly installed

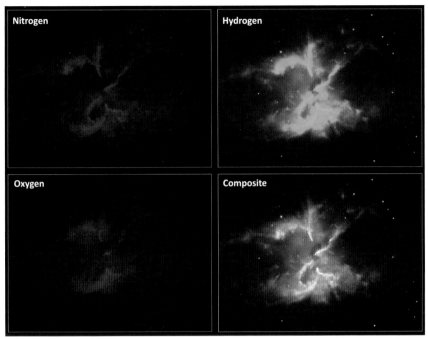

Figure 3.9-3. This color composite of the same imagery as in Figure 3.9-2 shows more distinct colors and shows the features within this object in greater detail. Here the hydrogen light is shown in green. (NASA image; credit: NASA, ESA, and the Hubble Heritage Team (AURA/STScI); http://heritage.stsci.edu/2009/05/.)

camera, the Wide Field Camera 3 (WFC3). This instrument incorporates a new generation of detector technology with greater resolution and sensitivity and includes a very broad selection of filters, similar to WFPC2. Here we are looking at five filters from the blue through the near infrared, which sample a fairly broad swath of the spectrum, unlike the previous example (Figures 3.9-1–3.9-3) where we were sampling light that was coming from a very specific color of light emitted by specific chemical elements. These filters are used to study astronomical objects in a different way; in particular they are able to distinguish various classes of stars by their colors and brightnesses. The WFC3 actually comprises two separate cameras, one sensitive to ultraviolet and visible light (WFC3/UVIS) and a separate camera sensitive to infrared light (WFC3/IR).

At the top of Figure 3.9-4 are the separate black and white images from each filter: from the left, a blue light image, a yellow-green light image, a hydrogen light image, and a red–near-infrared light image, all from the WFC3/UVIS camera, and finally one image (top row, right) from the infrared (WFC3/IR) camera. The challenge is how to put together these images into a single color composite. It was not as simple as assigning a filter to each additive primary, since there are more than three images. In general we assign the colors so that the reddest filter is assigned in a red color and the bluest filter is assigned in a blue color; the intermediate filters would be assigned in green or some intermediate color. Additional filters can be assigned a color intermediate to the primaries or more than one filter can be assigned the same color. In this example we have assigned red to both of the two reddest broad-band filters, red–near-infrared and infrared, as well as the narrow-band hydrogen light image. We have assigned green to the yellow-green filter and blue to the blue filter. The only somewhat nonintuitive technique is to shift the images of invisible infrared light into visible red.

While the choices of color may be somewhat arbitrary, the results are a real image and represent real physical processes that are occurring in these objects. You see real red areas in this image and real blue areas in this image, and those reflect real physical conditions that are going on in these objects. The colors are not made up, in other words. Some people have a misconception that we are assigning color arbitrarily or that we are painting these images just so they look spectacular and amazing. In fact, the color drops out

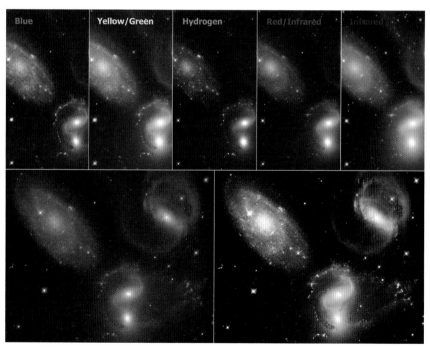

Figure 3.9-4. This slightly different example from Hubble's WFC3 is a bit more complicated because we had used five filters here instead of three. To visualize the data from more than three filters, we had to depart from the three-color model, and it required some fairly subjective choices. (NASA image; credit: NASA, ESA, and the SM4 ERO [Servicing Mission 4 Early Release Observations] Team; http://hubblesite.org/newscenter/archive/releases/2009/25/image/c/.)

of the data, and the colors are entirely representative of the physical processes that are going on within these objects. The bottom left panel of Figure 3.9-4 shows the initial color composite.

The other little twist is that the image at this stage is still a draft, and there are some techniques we can use to make the image a little bit more interesting visually. The bottom right panel of this image (Figure 3.9-4) is what the final result looks like. All we have done is adjust the contrast a little, adjust the brightness a little, tweak the color a little—the same kinds of things a photographer would do with the picture that comes out of his or her camera. The photographer might bring the image into Photoshop, twiddle a few dials, and make the image maybe a little bit more presentable, maybe something that will attract people and prompt them to look deeper to find out more about this object.

Most of these images are made to illustrate scientific findings from Hubble. Typically, the Space Telescope Science Institute (STScI) distributes a press release announcing a science finding by an astronomer and illustrates the main science points of that discovery with graphics. We want to make the images as presentable as possible so people will pay attention to them. We also want to draw people in to learn more about what the discovery is about or what Hubble is doing or what scientists are finding out because of Hubble.

The Hubble Heritage Project

With funding and encouragement from the Director's Office of the Space Telescope Science Institute, the overall methodology described above for creating and disseminating Hubble imagery as effectively as possible has been formally established as the Hubble Heritage Project, a small team based at STScI that consists of astronomers, multimedia artists, and outreach specialists. We use a two-pronged effort to find and distribute images from Hubble with the most visual impact, even though they may not necessarily represent breakthrough science. One approach is to search the Hubble data archive, looking through the store of science data that Hubble has accumulated over many years to try to find those images that can be made most attractive visually and aesthetically. We also have a *very* small fraction of Hubble's highly competitive observing time to take our own images. Most of these observations augment existing data from the archive, which by themselves may not be the best to produce a really pleasing result. With some additional data we can add a filter or another field to present a more complete image. In a few cases we have produced wholly new images from our observations.

The application of aesthetic principles to science observations is now common in astronomy, not just with Hubble but with ground-based observatories and the other Great Observatories. Other space-based observatories are also maximizing the return on their science investment by producing aesthetically rich images. These images are not only rich in science content but also in visual content and aesthetic content.[1]

A good example of this is how all three of NASA's Great Observatories have cooperated to create stunning images. Hubble along with the Spitzer Space Telescope, which operates in the infrared, and the Chandra X-ray Observatory, which operates in the X-ray part of the spectrum, observed the same region of the sky, the central portion of our galaxy, the Milky Way (Figure 3.9-5). The brightest part of the large main image (center right) is the actual center of our galaxy, and around it there is a lot of action going on, mainly star formation. There is crazy stuff going on there, all driven by a gigantic black hole at the galaxy's core. We took the three very different views that these three facilities provide and were able to put them together using the same paradigm—that method of combining the red, green, and blue components (panels below main image)—to produce a single image that represents all the observations from the infrared all the way to the X-ray. This is a mind-boggling concept to me, that in one picture you can see the entire span of the spectrum.

Figure 3.9-5. Separate and combined telescope images of the Milky Way. The brightest part of the image to the center right is the actual center of our galaxy. At top: the image composited by using separate datasets from NASA's Great Observatories. Bottom (from left): the separate images from Spitzer Space Telescope, Hubble Space Telescope, and Chandra X-ray Observatory Center (CXC) used for the composite image. (NASA image; credit: NASA, ESA, SSC [Spitzer Science Center], CXC, and STScI; http://hubblesite.org/newscenter/2009/28/.)

It is a bit of a challenge, because the images from these regimes can be qualitatively quite different. At the bottom of this image are the three separate images that went into producing the composite. You can see that they are visually very different. The most significant issue is the difference in resolution. As we have heard, the resolution of Hubble is better than with any other telescope, which means that we can distinguish very fine features in the images. The more finely resolved features in the Hubble image tend to disappear in the composite in favor of the broadened versions of similar features in the lower-resolution images (especially emphasized in a small reproduction). On the other hand, image features between the very different energy regimes tend to appear quite different in general. When these are combined in different colors, the features stand out quite clearly.

Producing these kinds of images for outreach purposes has become a very common approach now in astronomy and not only from the professional observatories. Because of a convergence of technologies, amateurs now are also producing amazing images from backyard telescopes. Very high-quality optics, capable detectors, and cameras are now much less expensive and more accessible to amateur astronomers than even in the recent past. In addition, high-performance computers and professional image-processing software are affordable and accessible. Nevertheless, it still takes a fair amount of preparation, skill, dedication, and sense of aesthetics to get the best results.

It may be a bit presumptuous, but I like to think that this convergence of forces permitting the flood of aesthetically pleasing images has pulled up the "production values" of publicly accessible astronomical imaging as a whole. There has also been a dedicated effort among several of the world's greatest observatories—ground-based and space-based—to not only publicize their science results but also to distribute aesthetically high-quality images along with a great deal of supporting, supplemental content. The result is a growing realization within the science community that such outreach efforts contribute to the awareness among not only the science-attentive segment of the population but also the general public and, perhaps most significantly, policy makers that astronomy and space science are valuable and important.

Note

1   Among many examples are David DeVorkin and Robert Smith, *Hubble: Imaging Space and Time*, with contributions by Elizabeth A. Kessler (Washington, D.C.: National Geographic, 2008; reprint, 2011); Edward Weiler, *Hubble: A Journey through Space and Time* (New York: Abrams, 2010).

10

Displaying the Beauty of the Truth: Hubble Images as Art and Science

Elizabeth A. Kessler

I do not remember the first time I saw a Hubble Space Telescope (Hubble) image. I wish I did. I wish I could tell a story about how I came upon the telescope's view of the Eagle Nebula or the Orion Nebula and felt overwhelmed by a deep sense of awe and wonder. Given that I have written a book about the Hubble images and their evocation of the sublime, it seems like a confession to admit this hole in my memory.[1] But although I do not remember a specific moment when I responded to the Hubble images in this way, I do see that power in them. It might be argued that their very subject matter leads to such a response—who is not awestruck by the stars? But the way the Hubble images portray the universe encourages and enhances this response. In many cases, the celestial scenes resemble landscapes of the American West, especially as depicted by nineteenth-century painters such as Thomas Moran and Albert Bierstadt. In their shape, color, and orientation, the pillars of the Eagle Nebula echo the rock formations in Moran's *Cliffs of the Upper Colorado River, Wyoming Territory* (Figures 3.10-1, 3.10-2). Because the Hubble images affect us this way and because they have circulated so widely, they have come to define how we imagine the cosmos.

Hubble has made great contributions to science, but those outside the scientific community are more likely to know and admire the Hubble's many images of nebulae, galaxies, and star fields. These pictures of the cosmos depend on the advanced technology of an orbiting telescope equipped with high-powered cameras and the careful choices of those who translate the Hubble's data into dramatic images. Many of the best known examples have been crafted by the members of the Hubble Heritage Project, a group of astronomers and image specialists at the Space Telescope Science Institute who have taken the monthly release of a new image as their mission. The Heritage Project images reach a broad audience, one that ranges from fellow scientists to school children. The group has a difficult task: how does one use an image to convey scientific information and to inspire an aesthetic response in a fash-

Figure 3.10-1. Popularly known as *Pillars of Creation*, this depiction of the Eagle Nebula is the most famous and likely most reproduced of all of those taken by Hubble. Its towers of interstellar dust and gas some 7,000 light years from Earth are nurseries of future stars, hence the name. Astronomers Jeff Hester and Paul Scowen crafted the image from Hubble data taken on 1 April 1995, and it was publically released in November 1995. In the years since, it has circulated widely in all manners of settings. (NASA image PR95-44A; http://grin.hq.nasa.gov/ABSTRACTS/GPN-2000-000987.html.)

ion that is legible to a group of people with widely varying understandings of the science, technology, and methods used to make the pictures? Not surprisingly, the resulting images are complex entities that raise questions about the place of images within science, their epistemological value, and the relationship of such images to a larger visual tradition.

It is that last issue that I will focus on here, in particular the question of whether Hubble images might be considered to be science or art. The Hubble images have characteristics we associate with both fields, and as a result they do not fit comfortably in either one. One response to this problem might be to introduce another term to describe such images.[2] After all, many images are neither art nor science but exhibit attributes connected to both. Another possibility is to think about how we define science and art. If one reflects on the response the Hubble images elicit, how they are made, and how they represent the cosmos, contemporary definitions of science and art seem unnecessarily narrow. This is hardly a new problem. Moran's paintings from the nineteenth century were criticized and praised for their affinities with

Figure 3.10-2. Thomas Moran, *Cliffs of the Upper Colorado River, Wyoming Territory,* 1882. (Smithsonian American Art Museum: Bequest of Henry Ward Ranger through the National Academy of Design.)

science, just as the Hubble images have been for theirs to art. It seems that the Hubble images invite us not only to look outward at the universe but also to reflect on the concepts we use to describe and categorize what we see.

I vividly remember the first time I saw an older set of astronomical images. Several decades ago the neon-bright false color photographs of Jupiter and Saturn from the Voyager mission were featured in a photography magazine that I brought to school for show-and-tell. But beyond the usual childhood trips to planetariums and an appreciation for a starry night sky, I did not have a strong interest in astronomy or even science in general when I was growing up. I was thrilled when I did well enough on my advanced placement exams to avoid college courses in science or math. I could pursue my interests in literature, art, philosophy, and even psychology (I overlooked the very central role statistics plays in this discipline) without distractions. The worlds of the humanities and science seemed very separate to me, even as the striking photos of distant planets remained in my memory.

Of course, I was far from alone in thinking this way. As C. P. Snow famously suggested more than 50 years ago, science and art can be imagined as two cultures.[3] Much of his essay reflects his historical era and a mid-twentieth-century belief in the power of science to solve the world's problems. But Snow's essay is not remarkable for how he characterizes the two disciplines; it is the idea of two divided cultures that continues to hold purchase today. It is easy to think of them as separate domains with distinct methods and understandings of the world, to imagine them as asking and answering different kind of questions, to believe that they share little with each other. To divide the fields in this way encourages a person to take sides (as Snow himself did) and to judge one or the other as the approach best able to serve humanity's interests.

But part of the appeal of the dichotomy is that it's endlessly enjoyable to break it down, to find moments of communion and connectedness between these two cultures, to identify ways in which one relies on the other, to point to common interests and concerns. The Hubble images might be seen as one such bridge between the two cultures. As the product of advanced scientific instruments that represent unfamiliar and even puzzling phenomena, their provenance and their subject matter label them as scientific images. And yet the appearance of the images and their appeal to our senses aligns them with art. In Hubble's portrait of the Eagle Nebula the columns of gas and dust reach up in dramatic fashion. Tightly framed by the Wide Field Planetary Camera 2's unusually shaped field of view and backlit by glowing gases, this small section of the much larger nebula gains an impressive monumentality. The Heritage Project's view of NGC 602, a nebula in the Small Magellanic Cloud, sparkles with a dazzling collection of stars (Figure 3.10-3). Curving layers of gas and dust pull our eyes into the depths of the nebula. There is no denying that these images engage our aesthetic sensibility.

For those involved with the Hubble Heritage Project, the aesthetic appeal of the images makes them akin to art. When I interviewed members of the group, I asked whether their images should be considered art or science. Howard Bond, one of the founders of the project, responded that he thought of the images as an "interface between them," and he went on to describe the goals of the project, saying that "it has to be a compelling image from the pictorial or artistic point of view. That's the number one criterion. On the other hand, I mean, there're certainly scientific implications, there're scientific processes that you can see going on in here, star formation. So really, it's a little bit of both."[4] Zolt Levay concurred, saying "I think the Heritage images are more about art.... The Heritage images are primarily about the visual.... I certainly hope that people are curious about them [the images] and want to learn what these things are...but I hope that they stand on their own visually."[5] Their responses imply that an image made solely for scientific purposes would have little aesthetic appeal, whereas the Heritage Project images are made with the intention of engaging our senses. The images convey information too, but this is not the first consideration when the members of the group choose a celestial object to observe or when they work to translate the

Figure 3.10-3. Star cluster NGC 602 in the Small Magellanic Cloud is depicted in an image captured with Hubble's Advanced Camera for Surveys in July 2004 and developed and released by the Hubble Heritage Project in January 2007. (NASA image; credit: NASA, ESA, and the Hubble Heritage Team [STScl/AURA]–ESA/Hubble Collaboration; http://hubblesite.org/newscenter/archive/releases/2007/04/image/a/.)

Hubble's data into visual form. Instead, aesthetic appeal takes priority.

The vividly colored, dramatically lit, and carefully composed images do not have to look as they do. As Zolt Levay explains in his essay in this volume, astronomers assign colors to exposures taken through different filters and combine these to create a composite with a full array of hues. To see the details in the nebulae and galaxies, astronomers also make adjustments to the contrast and boost subtle distinctions in tone, making them visible to our eyes. They can choose how to orient an image too. Cardinal directions have little significance for an orbiting telescope, and Hubble images often present the celestial scenes in a manner that maximizes their aesthetic appeal. In the Eagle Nebula, for example, north is diagonally to the left rather than at the top as convention might dictate.

It is the amount of careful crafting that has led others to suggest that the Hubble images should be considered art. Architect and sculptor Maya Lin, best known for her design of the Vietnam Veterans Memorial, writes

> [The Hubble Space Telescope] has shown us reaches of the universe that, until recently, were beyond comprehension. But the legibility of these far-off realms owes much to the engineers and scientists who interpret the telescope's imagery by specifically selecting the amazing colors we see. When I first saw NASA's renderings of distant nebulae, I realized that these weren't just reference photographs but, rather, works of art—and undoubtedly among the most powerful that our generation has produced.[6]

Lin recognizes the essential role of astronomers and acknowledges that it is their skillful interventions that elevate the Hubble images and transform them from scientific data into pictures that can elicit a strong emotional response. Although derived from scientific observations, these images express the awe and wonder that one experiences—whether an astronomer or not—when confronted by the vast size and scale of the cosmos.

In addition to appealing to the senses and exhibiting a high level of craftsmanship, the Hubble images also continue an established artistic tradition. In the nineteenth century artists such as Moran and Bierstadt traveled through the American West in search of the sublime experience and then painted spectacular scenes for an audience eager for a glimpse of the unfamiliar terrain. The artists used the visual vocabulary of romanticism to portray the towering peaks, deep canyons, and rugged rock formations. The Hubble images rely on a similar iconography to evoke the sublime. In the Eagle Nebula and many of Hubble's dramatic pictures of nebulae, the similarities are quite apparent. The profile of the columns recalls rocky buttes, signifiers of the American West. The color scheme, which creates yellowish brown pillars against a blue background, again looks like that of the western landscape. The details throughout the clouds give them the appearance of mass and substance. Nebula NGC 602 does not look like a landscape in the same way, but it shares an interest in conveying great size and scale with the nineteenth-century landscape paintings. In *The Chasm of the Colorado,* Moran took up the daunting task of framing the immensity of the Grand Canyon (Figure 3.10-4). He filled the canvas with a series of rock formations that extend to the horizon. Similarly, NGC 602 (Figure 3.10-3) depicts layers of gas and dust that extend into the distance.

Perhaps in recognition of these attributes, the Hubble images have been

Figure 3.10-4. Thomas Moran, *The Chasm of the Colorado*, 1873–1874. (Smithsonian American Art Museum, lent by the Department of the Interior Museum.)

exhibited at art museums and even have been acquired as part of museum collections. The Walters Art Museum in Baltimore featured several of the pictures in a special exhibit, integrating the Hubble images into the classical architecture of the building. Special prints of the Heritage Project images were made to fit into the arched insets.

Yet the Hubble images stand outside contemporary definitions of art. Today the concept behind a work of art often matters as much or more than the craftsmanship. Contemporary art explores the ugly and degraded with the intention of creating discomfort and uncertainty as often as it employs the beautiful or sublime to give pleasure. Originality and avant-gardism win high praise. When artists look back to older traditions it is often accompanied by a sense of irony rather than admiration, but it is possible to find exceptions: art that does not conform to these standards and is embraced by the tastemakers in the world of art.

More than a failure to fit contemporary notions of art, the Hubble images will not be widely accepted as art because they are made in the service of science. At one level this judgment is fair. The members of the Hubble Heritage Project are constrained by the conventions of astronomical representations. These are not images inspired by data; but rather, images that reflect the data. Again, Levay's essay (this volume) is helpful here. Colors have significance for those who know the key; they are not simply arbitrary assignments of different hues. But I suspect that there is also something prejudicial about the sense that a scientific image cannot be art, something based in the belief that the association limits (or even eliminates) the creative and expressive potential. The functional aspects of the images further disqualifies them because art is often imagined as something appreciated in and of itself without attention to what use value it may have.

The question of whether Hubble images are art would seem the more contentious side of the debate, whereas the question of whether they are science might seem a settled matter. But are they science? As Howard Bond proposed in response to my question about whether the Heritage Project images should be considered art or science, it is possible to see what he called scientific processes in the images. In his use of the phrase, scientific process means something like a natural process that we understand through the aid of science. However, discovery might be considered an essential component of science; to do science means to reveal something previously unknown about the natural world. It is this split between making scientific information visible and revealing new information that entangle the Hubble images. Every Hubble image illustrates details about the cosmos learned through the methods employed by scientists. But not every example shows astronomers something novel. In many instances, the visual expression is less enlightening than the numeric data than lies behind it.

Historian of science Peter Galison has described the ambivalent attitude of scientists toward images, and he aphoristically summarizes their position: "We must have images; we cannot have images."[7] Images appeal to the senses, which are easily tricked and deceived. Understanding numeric data requires the use of reason and analytic thought. It is a position often repeated by those who write about astronomical images.[8] Numeric data from Hubble allow astronomers to make calculations and determine the distance, magnitude, and other attributes of celestial objects. David Leckrone's discussion of magnitude demonstrates the level of precision that astronomers strive to achieve. An image may show that one star is brighter than another, but the eye cannot judge exactly how many time brighter.

The status of the Hubble images has also been called into question because they require astronomers to make the data visible. As Levay acknowledges earlier in this collection, some of the choices are inevitably subjective. Because infrared light lies beyond the range of human vision, Levay and his colleagues must decide how to represent it. For some this necessary intervention taints the scientific purity of the images.

But again, such a view depends on a very narrow view of science, one that ignores the numerous interventions that are necessary to gain knowledge of the cosmos as well as the shifting definition of what counts as an objective representation.[9] We cannot see the faint light of the nebula across a wide spectrum. Our position beneath the obscuring atmosphere of the earth and our weak eyesight make this impossible. With the aid of Hubble and digital image processing, as well as the expertise of those who use these technologies, it becomes possible to see such phenomena. The images are translations of numeric data into sensible form.

If we remain with the contemporary understandings of art and science, I find it difficult to resolve the question of the Hubble images' relationship to art or science. But the resemblance to nineteenth-century landscape representations offers an opportunity to think historically about their uneasy position. Few people today would argue that Thomas Moran's paintings of Yellowstone and the Grand Canyon are not art, and few would suggest that they should be considered science. *The Grand Canyon of the Yellowstone* and *The Chasm of the Colorado* today hang in the halls of the Smithsonian American Art Museum as impressive examples of American artists' celebration of the country's landscape. But the paintings came out of scientific expeditions. Moran visited Yellowstone in 1872 and the Grand Canyon the following year as part of government-sponsored scientific surveys. While his colleagues mapped the terrain, collected specimens, and documented the geology, Moran made sketches of the landscape, some of which illustrated official reports from the trip.[10]

The scientists on the survey team did not use Moran's paintings for their work, at least not directly. But the paintings *do* give us a great deal of scientific information about the Grand Canyon and Yellowstone, and this was acknowledged at the time. An anonymous art critic in *Scribner's Monthly* wrote of Moran's *The Chasm of the Colorado*: "It is not paint that one sees; it is a description so accurate that a geologist need not go to Arizona to study the formation. This is geology and topography."[11] Moran himself insisted that it was a particular understanding of the place that allowed him to paint it, writing later in his life that "[i]n condensed form, this is my theory of art. In painting the Grand Canyon of the Colorado and its wonderful color scheme...I have to be full of my subject. I have to have knowledge. I must know the geology. I must know the rocks and the trees and the atmospheres and the mountain torrents and the birds that fly in the blue ether above me."[12] This is the very type of knowledge that his fellow travelers on the survey would be acquiring, and one assumes that Moran gained greatly from his conversations with them.

In a confirmation of the success of Moran's method, John Wesley Powell, a well-respected scientist and the head of the survey through the Grand Canyon, responded to *The Chasm of the Colorado* by writing that "it required a bold hand to wield the brush for such a subject. Mr. Moran has represented depths and magnitudes and distances and forms and colors and clouds with the greatest fidelity. But his picture not only tells the truth, it displays the beauty of the truth."[13] Powell praised not only the accuracy of the painting, but he also identified an alliance between aesthetics and truth. To tell the truth pictorially is to appeal to the mind, which can analyze the faithfulness of the representation to the original scene. But to display the truth—to exhibit it—is to expect a response of the senses and the body, a recognition that goes beyond merely checking the facts. When coupled together, these two modes of expression give a more complete picture of the Grand Canyon. To ignore the aesthetic experience would deliver only a half truth.

To return to the Hubble images, I think it is appropriate to rephrase Powell and suggest that the Hubble images not only tell us the truth about the cosmos, but they also display the sublimity of that truth. For many in the humanities today truth is an elusive concept, determined by culture not by some universal rule. Scientists more readily embrace the possibility of finding truth, at least in terms of gaining a certain understanding of the laws of nature. Powell's words make clear that aesthetics can aid in that quest. The Hubble images allow us not only to know facts about the cosmos but to experience them as well.

Although I do not remember when I first saw them, I can say with confidence that my interest in the Hubble images arose from their aesthetic appeal. They reminded me of landscape paintings and the artistic tradition of romanticism, a genre and an artistic movement that I found fascinating. I was naïve when I began researching and writing about them; I did not consider how much science I should have to learn to understand and to write effectively about astronomical images. It was fortunate that I did not reflect too much on this as I would have been daunted by the prospect. In the end, studying Hubble's images required something else of me: to reconsider my assumptions about how we define art and science.

Notes

1   Elizabeth A. Kessler, *Picturing the Cosmos: Hubble Space Telescope Images and the Astronomical Sublime* (Minneapolis: University of Minnesota Press, 2012). The book discusses in more detail much of what is introduced in this essay.

2   This is the approach that art historian James Elkins takes in *The Domain of Images* (Ithaca: Cornell University Press, 2001).

3   C. P. Snow, *The Two Cultures* (Cambridge, Mass.: Cambridge University Press, 1998).

4   Howard Bond, Oral History Interview by E. A. Kessler, 16 September 2003, Kessler personal notes, 18.

5   Zoltan Levay, Oral History Interview by E. A. Kessler, 14 October 2003, Kessler personal notes, 28.

6   Maya Lin, Artforum (http://artforum.com/inprint/id=30066).

7   Peter Galison, "Images Scatter into Data, Data Gather into Images," in *Iconoclash*, ed. Bruno Latour and Peter Weibel, (Cambridge, Mass.: MIT Press, 2002), 300.

8   See James Elkins, *Six Stories at the Edge of Representation: Images in Painting, Photography, Astronomy, Microscopy, Particle Physics, and Quantum Mechanics, 1980–2000* (Stanford, Calif.: Stanford University Press, 2008); also Michael Lynch and Samuel Y. Edgerton Jr., "Aesthetics and Digital Image Processing: Representational Craft in Contemporary Astronomy," in *Picturing Power: Visual Depiction and Social Relations*, ed. Gordon Fyfe and John Law, Sociological Review Monograph 35 (London: Routledge, 1988), 184–221.

9   On the history of different understandings of objectivity, see Lorraine Daston and Peter Galison, *Objectivity* (New York: Zone Books, 2008).

10 On Moran's relationship to the scientific surveys, see Joni Louise Kinsey, *Thomas Moran and the Surveying of the American West* (Washington, D.C.: Smithsonian Institution Press, 1992).

11 "Culture and Progress: 'The Chasm of the Colorado,'" *Scribner's Monthly* 8 (July 1874):374.

12 Thomas Moran, "Knowledge a Prime Requisite in Art," *Brush and Pencil* 12 (1 April 1903):14.

13 Quoted in Kinsey, *Thomas Moran,* 112.

Epilogue: Exhibiting the Hubble Space Telescope

David H. DeVorkin

The essays in this book provide personal insights from participants who dreamt of a space telescope, built a space telescope, used one, and explored its significance to science, art, and history. These essays are packed with personal information based upon experience and laced with emotion and pride; they are, in many respects, personal journeys. As structured testimony, they reveal the deep impact the Hubble Space Telescope (HST) has had on careers and lives. And as participant history, they provide perspectives that in a few decades will no longer be available.[1]

The editors of this volume and two of the three section moderators are, or have been, associated with efforts to preserve and present the history of the HST. Their efforts have been cited throughout this volume and are well known in the community. At another level, three of them (DeVorkin, Smith, and Tatarewicz) have been involved in bringing the promise and the products of the HST to the museum-going public and placing them in context with respect to spaceflight and culture. The choices they made in the process since 1981, and the results, illustrate how the National Air and Space Museum (NASM) wished to portray its legacy to the millions of visitors passing through its halls. So as an epilogue to this volume, building on Tatarewicz's description of the Structural and Dynamic Test Vehicle in his introduction to Part 2, we recount the many ways we have presented the HST (originally the LST, or Large Space Telescope; then ST, or the Space Telescope; finally HST, for Hubble Space Telescope, since 1983) and astronomy to the public since the early 1980s.

When NASM opened in 1976 with 23 galleries and 3 major halls devoted to the history and celebration of air and space flight, there was little contextual coverage of space astronomy. The Princeton University Optical Telescope Assembly (OTA) from the Orbiting Astronomical Observatory (OAO) 3, Copernicus, was displayed near a case containing a drop test model of "Little Boy," the atomic bomb dropped on Hiroshima, within a gallery originally titled "Earthbound Benefits from Flight." They both were cylinders. There was a gallery on "Life in the Universe" that explored the question fancifully, but there was no portrayal per se of space astronomy.

The first object brought into the collection that depicted the ST was a one-fifteenth scale model of the satellite, provided by the Lockheed Corporation, to illustrate a space transportation exhibit in the spring of 1981. The model eventually became part of a prominent display "America's Space Truck, the Space Shuttle" that portrayed the ST as a payload for the Space Shuttle.[2] Thus in its first five years of life, the NASM devoted little if any space to the prospect of a space telescope. By the late 1990s and through to the present, this would change to the point where more floor space is now devoted to the HST, its elements, and its products than to any other single space mission with the exception of the Apollo series, Skylab, and the Space Shuttle.

The Stars Gallery 1983–1997

"Stars: From Stonehenge to the Space Telescope" opened in 1983 featuring, as illustration of its title, a one-fifth-scale model of the HST, donated by Lockheed, hanging in space above a walk-thru mock-up of a Stonehenge arch (Figure E-1). Clearly our intention was to display the HST as a critical milestone in astronomy. There were only identity labels in the entrance, and the

Figure E-1. The "Stars: From Stonehenge to the Space Telescope" gallery opened in 1983, just as the mission was being renamed for Edwin Hubble. At the opening (left) visitors were treated to a vision of an HST servicing mission (right) featuring a one-fifth-scale model of HST donated by Lockheed Corporation. (Author's photographs.)

one for HST merely stated it would be launched by the Space Shuttle in 1985.

Deeper into the gallery, a 5,000-square-foot affair that explored how we study the Sun as a star and the stars as suns from the ground and in space, we placed a one-fifth-scale model of the OTA for the HST in a case that outlined the spacecraft. Donated by Perkin-Elmer Corporation, this model boasted high definition mock-ups of the optics and truss (Figure E-2).

The exhibition strategy here was to use the HST model to display how we observe the universe in the visual and ultraviolet (UV) regions of the spectrum from a vantage point in space. The entire back wall and part of one side wall was a huge graphic of the electromagnetic spectrum, with engineering models and mock-ups of representative missions. In order of descending energy detection range (or increasing wavelength detection range), we had a full-scale engineering model of Uhuru, then a full-scale engineering mock-up of the International Ultraviolet Explorer (IUE), then the flight prototype of Princeton's OAO Copernicus, then the HST scale model, and, for the infrared, Frank Low's original bolometer in a mock-up of the National Aeronautics and Space Administration's (NASA's) Lear Jet fuselage. Several years later we added a reconstructed mock-up of the Infra-Red Astronomy Satellite (IRAS) and moved the Caltech 2.2 micrometer survey telescope from another part of the gallery to illustrate the ways that infrared astronomy was established as a mainstream specialty in the USA.

Figure E-2. Perkin-Elmer Corporation donated a one-fifth-scale model of the HST's Optical Telescope Assembly, which was featured in the "Stars" gallery along with descriptions of the primary instruments. There was also a timeline that needed revision after the fatal loss of *Challenger* in 1986. (Smithsonian image by NASM staff photographer.)

An introductory panel to this major section promoted the advantages astronomers had viewing space from above the Earth's atmosphere. "Why Go into Space to Do Astronomy?" oriented the visitor to how the full electromagnetic spectrum would be made accessible with no obscuration or blurring of images. There was a mechanical interactive allowing visitors to move a piece of semi-transparent blue plastic in front of a star field. The point was made, however, that "ground-based and space-based astronomy work together today" because larger optical and radio apertures were economically feasible on earth.

Each of the displays for the satellites described how they worked and what they observed. Images of celestial objects produced by the instrument, along with graphics of their optical systems, provided hints of both product and process. In the high-energy realm, these images came from the suite of High Energy Astrophysics Observatories. The IUE display emphasized how a space telescope in orbit could be designed to make the observing experience directly available to astronomers, as if it were a telescope operating in the next room. It was presented as heralding a new era of observational astronomy in which space observations could be taken as directly as ground-based observations and astronomers could competitively bid for observing time through a peer-review process, as was the case for the national observatories. Up to the time of the IUE, access to space astronomy missions was primarily through the teams and institutions who built the payloads. The OAO 3 developed a small visitor observer program as a trial run, but IUE was designed to be used that way from the beginning. Although there was a detailed description of this capability in the IUE panels in the exhibit, the matter was not addressed by the HST panels.

Beyond the IUE was the Princeton OTA from Copernicus, a full-scale operational prototype of the 32-inch (-0.8 m) reflector equipped with a far UV spectrometer. The prototype was suspended at an angle from the ceiling, and its interior was illuminated to reveal the optics. The featured artifact from the mission was a high-work-function open-cathode photomultiplier that was in the original set of detectors that Princeton delivered to NASA for the flight and was part of the testing process. The point of displaying this object was to highlight the design changes necessary to make the instrument sensitive to the far UV portion of the spectrum. So once again, it was a display of the technical modifications needed to conduct astronomical observations from space.

In hindsight, the placement of the full-scale IUE and the OAO artifacts just prior to the presentation of the HST through the one-fifth-scale models was problematic. It mixed scales, which can be misleading, but more important, it strongly implied that there was a natural progression to the HST, from a 16-inch (-0.4 m) reflector in geosynchronous orbit, to a 32-inch (-0.8 m) reflector in near-earth orbit, and finally to a 94-inch (-2.4 m) telescope. In fact, if a visitor carefully read the labels they would know that the IUE was launched in 1979 and the OAO in 1972, thus reversing this apparent trend. But in exhibitry of this type, one must be more sensitive to the impact of the large objects and the likelihood that few if any of the visiting public actually read the labels. In any event, there was nothing in the labels themselves that intentionally indicated a progression, either in the manner in which the telescope would be employed (i.e., in the manner of IUE) or in the size of the telescope. Nevertheless, they set the stage for the HST.

In an early 1983 version of the script, "The Space Telescope," the OTA was fully described, stating only that it "will look deeper into space than ever before." A descriptive panel described how it would be launched into orbit by the Space Shuttle, and then the rest of the treatment briefly illustrated stages in its construction and its major components, starting with the mirror and then the scientific instruments. A large label and graphic described "How the Space Telescope Sees Objects," highlighting charge-coupled devices (CCDs), photomultipliers, and the radio link to Goddard Space Flight Center and the Space Telescope Science Institute.

With the fatal loss of *Challenger* in January 1986, we realized that we had to remove any reference in the gallery to a specific launch date for the HST. Actually we well knew it was needed because we still had 1985 on the wall, indicating that the confidence we had in a launch date when the gallery opened had evaporated by then. This was not yet critical until the *Challenger* disaster and the resulting halt in shuttle flights made it impossible to state any launch date. Rectification was necessary and was made possible by support we received from Ball Aerospace and Technologies Corporation and from NASA to upgrade the infrared section by adding the IRAS satellite mock-up. Then, more critically, we realized that the simplistic descriptions in the original 1983 panels could be enriched by the historical scholarship then underway at NASM in the form of the "Space Telescope History Project." This was a multi-institutional initiative between NASM and The Johns Hopkins University to document the conception, selling, and building of the HST. Paul Hanle of NASM and Robert Kargon of Johns Hopkins secured NASA funding for this project, hiring Robert Smith and later Joe Tatarewicz to organize and execute the historical project, which included conducting extensive oral histories, discovering and organizing historical records, organizing visual resource materials, and, eventually, creating a fully documented scholarly history of the telescope. More like "combat history" than what historians traditionally had tackled, the result was an award-winning book,[3] and numerous other products.

By the spring of 1986, Joe Tatarewicz produced a new script for the back wall section that included a detailed timeline describing the history of conceiving, lobbying for, and finally defining what was now called the Hubble Space Telescope. The new display panel declared "The Coming of Space Telescope," but the timeline made it clear that the dream just didn't happen. As the label text explained, "Between dreams and their realization lay a long, rugged road of development." The timeline included the images of past U.S. presidents, to illustrate historical context, along with photographs, graphics, and specimens, starting with Herrmann Oberth's 1923 book, *The Rocket in Planetary Space*" (*Die Rakete zu den Planetenräumen*), in which he "speculated on the advantage a large telescope would have if placed in space...where the stars do not twinkle." The next stop was Lyman Spitzer's 1946 Air Force Project RAND (Research and Development) study, "Astronomical Advantages of an Extraterrestrial Observatory," introducing the astronomer who more than any other lobbied for the telescope. Other stops along the way identified the National Academy of Sciences Space Science Board study between 1965 and 1969, the Large Space Telescope Steering Committee of the early 1970s, a series of "telescopes that never were," illustrating a 50-inch telescope that the Association of Universities for Research in Astronomy (AURA), in collaboration with the Army Ballistic Missile Agency, envisioned for Von Braun's space station. These and others through the 1960s included human-tended scenarios and photographic retrieval, including a proposal from the Langley Research Center to utilize Mercury hardware to create a returnable telescope. Labels described the problems of physical retrieval but did not dwell on the limitations of electronic image detection at the time. One highlight, centered on the time period between 1969 and 1973, was how the Large Space Telescope was originally to be launched on a Titan III and was switched to the proposed Space Shuttle. Although the 1975 reduction of the mirror size from 3 meters to 2.4 was on the time line, there was no elaboration as to why this was done. Other entries included background on Edwin Hubble and why the telescope was named in his memory, stages in the construction and assembly of the telescope at Perkin-Elmer and then Lockheed, astronaut servicing training, the establishment of the Space Telescope Science Institute in 1983, the first call for proposals in 1985, and full testing of the flight instrument in 1986 subsequent to the *Challenger* disaster. We placed a small audiovisual unit at the end of the timeline to provide a variety of updatable short clips, including commentary by Robert Smith. The label read "watch this screen for recent information on the activities of the Hubble Space Telescope."

Descriptive labels for the OTA did not change, nor did descriptions of the operation of the telescope. The museum also decided that the labels and general coverage would not have to be substantially changed once the telescope was launched, unless another disaster occurred. So overall the treatment was linear and clinical. Visitors would learn that the telescope had long been a dream of astronomers, it was reduced moderately in size, it was to be launched and serviced by the Space Shuttle, and it was to be available to all astronomers in a manner that had been proven by the success of the IUE. The HST's history was, in other words, linear and not problematic.

These first efforts in the 1980s did not extend to the display of actual HST hardware, nor did they encompass what have become larger issues in the history and life of the HST. Nothing was said about the military role in large orbiting optical imagers; servicing was taken as a fact, as well as the downsizing of the mirror; and the various purposes for which the HST was being built, the questions it would address and hopefully answer, were not primary issues that curators thought were proper or effective for exhibit. No one anticipated the huge impact the imaging would make on the press and public, for instance, nor the end of the Cold War and the declassification of a portion of the U.S. reconnaissance satellite program.

The Structural Dynamic Test Vehicle

The two one-fifth-scale models described above were acquired in 1983. The next object was the first we acquired that actually had played a role in the development of the HST itself, the Structural Dynamic Test Vehicle (SDTV). Lockheed built a full-scale engineering mock-up in 1975 for feasibility testing. Initially it was a low-fidelity metal cylinder with a base some 168 inches in diameter, an upper section depicting the telescope tube some 120 inches in diameter, and an overall length of 508 inches. The mock-up was continuously modified as Lockheed proceeded through feasibility studies and was finally awarded the contract to build the actual spacecraft. Among other tasks, the mock-up served as a frame on which the cables and wiring harnesses for the actual spacecraft were fabricated. It was also used for simulations in developing maintenance and repair activities in orbit. Dynamic studies on the test vehicle including vibration studies and thermal studies led to its being designated the HST SDTV.

When its useful lifetime ended, the object was stored outdoors at Lockheed in Sunnyvale, California, until it was donated to NASM in June 1987 and shipped to the Paul E. Garber Restoration and Storage Facility. There it was refurbished and restored to its 1976 configuration and went on display in the Space Hall in March 1989.

Exhibitry surrounding the SDTV in 1989 included a series of labels, graphics, and photographic panels (Figure E-3). These documented its use as a test vehicle, how it served as a wire-form to develop the cabling and wiring harnesses, and how it acted as a stand-in for photographs used by Lockheed to propose building the flight artifact. It simulated all known handling procedures for the flight artifact. In proposing that NASM acquire this object the curator, Joe Tatarewicz, pointed out, "Because of the size and cost of the observatory, only one flight spacecraft was built, and few spare parts are being procured." His proposal also stated, "Reassembly and refurbishment to flight appearance should be relatively easy. Such refurbishment would do no real damage to the artifact."[4]

With acquisition, plans for exhibition changed from simulating flight appearance to exhibiting the object as it was during its operational lifetime. "First, the visitor must not confuse the HST/SDTV with the real HST, and must not think that the real HST looks like the SDTV or behaves exactly like it. The exhibitry will identify the artifact for what it really is." The point of view of the exhibit was to "explain the difference between prototypes and protoflight" test objects and to describe the developmental life of the SDTV. There would also be a series of images depicting how the object changed in appearance as its function evolved.[5]

The exhibit that opened in 1989 detailed these themes, describing its

Figure E-3. The Structural Dynamic Test Vehicle was first displayed only with its core test components (left) but in the 1990s was upgraded to simulate flight appearance (right). (Left, author's photo; right, Smithsonian Institution photo by Eric Long, NASM.)

components, detailing its operational lifetime functions, and providing background on the many technical challenges facing the HST mission, including its suitability for servicing. No mention was made of its relationship to reconnaissance programs other than a single oblique reference in one of the photograph captions of a scene at Lockheed, Sunnyvale: "The structure in the background is an assembly stand used for other spacecraft that served as a model for the Space Telescope's stand."[6]

By 1996, with the HST a functioning observatory, the museum decided it was more helpful to portray SDTV as the HST in its operational mode since the museum had learned in the interim that indeed the SDTV had been employed to fit out the flight coverings. The SDTV was duly removed from exhibit in order to upgrade the object to simulate the actual HST as it was being deployed from the Space Shuttle. This major upgrade, generously supported and executed by Lockheed Martin Missiles and Space, Lockheed Martin Technical Operations, and HST subcontractors (Jackson and Tull, NSI Technology Services, Swales & Associates, and Hughes Danbury Optical Systems), and NASA's Goddard Space Flight Center, working with NASM staff and volunteers, involved fabricating the equipment section for the OTA and adding an aperture door, high-gain antennae, solar arrays, aft shroud handrails, and numerous other nonfunctional details. To make it appear to be in flight-ready condition, realistic multilayer (nonflight) thermal blanketing and taping, interface hardware, wave guides, and the umbilical were added. Lockheed and NASA also provided a large equipment cradle to allow the upgraded object to be displayed from the floor at a dramatic angle. The completed artifact was reinstalled in Space Hall in early 1997.

Labeling around the upgraded mock-up was duly revised not to describe the actual object on display but to celebrate the history of the spacecraft in orbit through the mid-1990s, which included its launch in 1990, the discovery of the flaw in the optical system, and the servicing mission in 1993 that saved the HST. The decision to upgrade the exhibitry, including the artifact itself, was not without some concerns. There was some sentiment that we should not conduct this alteration to make the SDTV look like the flight model. However, when we were advised by NASA and Lockheed that at times much of the flight hardware was tested using the SDTV as a framework, including the blanketing, we decided it was sufficiently within the use history of the object to make the changes. Thus, on the one hand, one might criticize our effort, which resulted in the transformation of an artifact into an icon, however on the other, everything that was done is fully reversible, and we created a new and powerful focus for what was, by then, a scientific spectacular.

The original placement of the SDTV in Space Hall was based upon its size and where it could be placed. There was no thematic context in Space Hall at that time. However, its removal from Space Hall coincided with a general recontextualization of the area by curators in the Space History Division. Their intention was to place the rockets, missiles, and spacecraft already in the hall into a meaningful Cold War context to help visitors appreciate why these objects were built in the first place: paid for by the imperative that we called the "Space Race." As we were planning this effort, essentially an extensive relabeling, we were provided a wonderful opportunity to borrow a collection of Soviet craft, space suits, and paraphernalia that had been recently purchased at auction by Ross Perot. Thus for the first time we were able to present aspects of both sides of the race into space—for the delivery of nuclear warheads, the human occupation of space, the sending of humans to the Moon, and military reconnaissance. This last aspect of the "Space Race" was highlighted by the display of a recently acquired Corona reconnaissance satellite, an acquisition made possible by President Clinton's declassification of the program and announced by Al Gore in February 1995.[7] When the refurbished HST was brought back into the hall, it was consciously placed at the end of the military reconnaissance section. In a 1993 treatment and "first cut," curators depicted reconnaissance in space as "a convergence of the arms and space races," suggesting they would pair a Salyut 5 film return canister with the American Discoverer 13 film canister, as well as display the "Hubble Space Telescope interpreted as a KH-11."[8] But after we knew that we would acquire an example of a KH-4 system, known as Corona, we made the inference that the HST was "Corona's cousin." A NASA reviewer of the script objected to this link, stating, "This is sheer speculation and I suggest sticking to the facts."[9] Indeed, we could only speculate on the connection but thought the connection had been clear enough and, indeed, had been commonly noted in the literature. My personal recollection is that one of the national networks used a graphic image of the HST, pointing down, to describe military reconnaissance from space vehicles! Nevertheless, although we refrained from any familial connection in the labeling, the HST stayed put at the end of the reconnaissance section, between Corona and the HST's transport vehicle, a scale model of the Space Shuttle.

The Faint Object Spectrograph

Within two years, we acquired the first major flight instrument from the HST. After the February 1997 servicing mission, when the Faint Object Spectrograph (FOS) was removed from the HST by the Shuttle astronauts and brought back to Goddard, we worked with Goddard to remove some of

The Structural Dynamic Test Vehicle

Joseph N. Tatarewicz

Identifying, locating, acquiring, and restoring the SDTV that was on display in NASM's Space Hall from 1989 to 1996, before it was removed for the upgrade, was a curator's version of "big science" and, as befits its association with the HST, also a grand challenge. As Smithsonian curators scoured industry for components, documents, and knowledge, we kept seeing pictures of something that looked halfway like the HST but in important ways was rather different. Everybody told me there was only one spacecraft, that there was never a prototype. A vehicle called "the mock-up" turned out upon further investigation to be an engineering test vehicle that responded to acoustic and various other vibration inputs much like the real spacecraft and was used to design some of its fundamental structural characteristics. It was also, it turned out, the structural and dynamic test vehicle to another set of satellites used for national security, which everyone declined to identify and got very uncomfortable whenever conversation turned to that subject.

In addition to the so-called modal design and testing, the vehicle was also configured cosmetically to resemble the 1970s design for the telescope, weighted and balanced with ballast, and used to determine and verify how the flight spacecraft would be handled on the ground. The 15 miles of flight wiring harnesses for the actual spacecraft were fabricated on the test vehicle and then transferred over. More important, however, and to the point of this section, was test vehicle's role in developing tools and procedures for on-orbit servicing.

As with so many of the actual scientists, engineers, and administrators associated with the HST, neither I nor my colleagues at the museum could have imagined that this mission would turn into such an intimate experience, a lifetime association. After we identified and acquired the HST's full-scale SDTV, we took great pains to exhibit it faithfully as it was employed in life to develop tools and procedures for on-orbit servicing. The display even served as a public demonstration of the servicing function when astronauts visited and were hoisted up to practice their art.

The decision to collect, preserve, and display the SDTV in our museum's Space Hall required careful planning and considerable convincing that the effort would be worth the cost in manpower and real estate. But it has proven to be a highly effective and constant reminder of the magnitude of the mission and has become the focus for the display of instruments returned from the last servicing missions as well as the iconic images gathered by those instruments and their successors. Some of my colleagues at the museum resisted acquiring this gargantuan artifact and had to be convinced that it was a genuine, historic piece of engineering and not "just a mock-up." Some worried about the work, expense, and disruption associated with inserting such a vehicle into the already crowded Space Hall, especially since the HST was expected to have only a nominal 10-year working life.

the FOS's interior insulation so that its digicon detectors and other optical elements could be rendered visible to our visitors. We constructed a small display centered on the FOS that highlighted the confirming evidence it provided for the existence of a supermassive black hole in the core of the giant elliptical galaxy M87 as well as the spectral signature of the impact of Comet Shoemaker–Levy 9 on Jupiter.

The Need for a New Exhibition to Replace Stars

In 1997 the old "Stars" gallery was closed to prepare for a new exhibition entitled "Explore the Universe." The fifth-scale models were removed and placed in storage, the OTA was soon loaned out to another museum for display, but the satellite model was slated for the new gallery. By now, due to the happy fact that servicing missions were returning actual flown instruments to Earth and our deepened appreciation, through the continuing research by Robert Smith and Joe Tatarewicz, for the history behind selling and building the HST, a more sophisticated treatment was called for. However, there was no linear path to this process. Exhibit development was highly contingency driven, as a brief look at the history of both the "Stars" gallery and its successor will attest.

"Stars" was originally initiated as a NASM priority sometime in 1980 by the museum director at the time, Noel Hinners, a former NASA administrator whose background was in science. Initially a series of curators and scientists deliberated over what it should include, and finally by mid-1981 the author (DeVorkin) assumed sole curatorial direction. The issue at hand for this exhibition, which was conceived, built, and opened solely on federal funds— common prior to the Reagan presidency but rare in its wake—was how to create an exhibition that was consonant with the overall philosophy of NASM. It was not the first gallery built on a space science theme, but it became a playing field for conflicting values over how best to serve the mission of the museum, as distinct from the Smithsonian. Especially after Hinners departed in the spring of 1982, the future of his gallery came into question by the new administration, worried that another gallery about "science" would further dilute the museum's message. Means to "humanize" the message of the exhibition, inspired by the new director, included an introductory section on the images of the Sun and stars in all cultures. There were also other suggestions made by museum administrative staff that, while in and of themselves were interesting, led to something of a hodgepodge. Although there were veiled threats from staff that local sports fans would inflict damage to a display case if we elected to include a Dallas Cowboys helmet (in Redskins territory), more seriously there was definite resistance on the staff against balancing the display between ground and space-based astronomy and, most of all, presenting science on its own. After all, we were an air and space museum. Thus there was little question that objects such as the Apollo Telescope Mount, the Orbiting Solar Observatory 1, Uhuru, IUE, Copernicus, or the HST would be included and celebrated as missions. In fact, this was the first time the HST was exhibited as a scientific instrument rather than just as a payload for the Space Shuttle. But there was concern that a significant amount of floor space was turned over to terrestrial telescopes, even if they opened new spectral regions for astronomy, like the Caltech 2.2 micrometer telescope or a model of the National Radio Astronomy Observatory 140-foot (-43 m) equatorial radio telescope. Even so, the exhibition opened with good press and long lines. At one of the openings, both the director and deputy director privately expressed mild surprise that it was so popular.

"Stars" remained popular with the public through the 1980s, but when a new director came in 1987, the astrophysicist Martin Harwit, he found it intellectually wanting. At the same time, a new chief of exhibits was hired, Nadia Makovenyi, who felt the gallery lacked a design coherency or any sense of celestial wonder and awe. Coincidentally, DeVorkin had wanted to make small improvements to the gallery, but by 1989 Harwit called for a completely new theme and treatment: "What do we know and how do we know it" was his directive to the staff. DeVorkin was delighted but soon found that the new gallery would be the combined responsibility not only of the curatorial division but of a new group of infrared astronomers Harwit had brought to the museum as a contingency of his hiring.

Coincidentally, as the gallery became a joint effort of the two departments, DeVorkin secured a sabbatical to finish up a long-standing biography. Robert Smith joined with Matt Greenhouse, Howard Smith, and Jeff Goldstein of Harwit's Laboratory for Astrophysics, and by 1992 they were joined by Valerie Neal as curatorial coordinator along with Beatrice Mowry as the designer and David Romanowski as editor. Through the first half of the 1990s and the launch of the HST, its subsequent troubles, and its (and NASA's) rescue, this team deliberated over many options, and the HST took center stage. One version of the planning document envisioned our visitors entering the gallery through a full-scale mock-up of the OTA, becoming, or assuming as it were, the role of a photon reflecting off the mirrors, being collected by detectors, and thereby revealing the wonders of the universe. By the time Harwit left the directorship in 1995, the theme of the gallery had focused on the history of observational cosmology, and Robert Smith had already identified some key artifacts, including the availability of William Herschel's original 20-foot reflector and the possible accession of the backup primary mirror from the HST, in storage at the Perkin-Elmer Corporation in Danbury, Connecticut.

With Harwit's departure, the subsequent closure of the Laboratory for Astrophysics, and then Robert Smith's departure for academia, DeVorkin stepped back into a curatorial role and with Mowry and Romanowski created a core team to build upon what Smith had established. The general historical thread was set as "New Eyes, New Universes," emphasizing observational cosmology, tracing how our conception of the universe changes as our tools of perception change. We started with visual sightings aided by pointing devices and then came the telescope, then photography, then spectroscopy, and then digital detection. This final addition, along with access to space as well as continued ground-based efforts, led to a vast expansion of the energy spectrum available. The revolutions that accompanied each change in perception are, in turn, heliocentric, galactocentric, acentric, and the dark universe.

Over 60 percent of the gallery is given over to the "The Digital Universe," which is effectively post-1960. The centerpiece of that section is the HST backup mirror, in a display titled "Collecting 'Core Samples' of the Universe" highlighting the Hubble Deep Field. Rather than displaying the HST as payload, as we did in 1981, or describing its timeline and anatomy, as we did in 1983 and 1986, now we emphasized the science that was done to date and the instruments responsible for that science. The Hubble Deep Field image is center stage on the panel, describing how it was created and the milestone confirmation "that galaxies began as small, irregular clumps of matter that merged into ever-larger clumps and eventually formed large, well-defined elliptical and spiral galaxies."[10] Other images on the panel extended this discussion with images from the *Near Infrared Camera and Multi-Object Spectrometer* secured in 1998. But the central story remained on the Hubble Deep Field and the history of the instrument that produced it.

We had acquired two complete optical channels from the Wide Field Planetary Camera (WFPC) instrument returned from the first servicing mission, and the Goddard Space Flight Center inserted them into an engineering mockup of the body of the camera. (Figure E-4) This arrangement nicely illustrates how the light entered the instrument and was split by a pyramidal mirror onto a pair of folding mirrors and finally into a small telescopes housing CCD detectors. This display was set behind a developmental history of the WFPC, illustrated by the original beam splitter from Jim Gunn's "four-shooter" camera from Palomar under the title "The 4-Shooter: A Test of the WF/PC Concept." This made the point that at the time of building the camera and choosing a detector, CCDs were believed to be more reliable than secondary electron conduction vidicons, but as yet they could not cover suitably large areas and so several had to be used together in a mosaic. The problem we wanted to illustrate, of course, using the four-shooter pyramid was that given the state of electronics

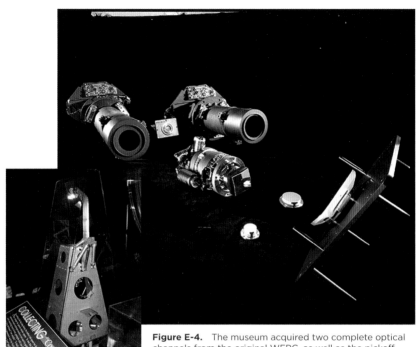

Figure E-4. The museum acquired two complete optical channels from the original WFPC, as well as the pickoff mirror, the beam splitting pyramid, and folding mirrors, all returned after the first servicing mission. These were fully documented and mounted in an engineering mock-up of the optical assembly, now on view in "Explore the Universe." (Smithsonian photograph by Eric Long, NASM; inset, author's photograph.)

at the time, these mosaics could not be continuous, and so a step toward the WFPC design was to try it out on a ground-based telescope.[11]

We chose to display the HST backup mirror prominently as the defining artifact for the section. Mounted vertically on a transport and testing stand, we introduced it as "one of two nearly identical main mirrors built by Corning for the Hubble Space Telescope" (Figure E-5).

Beyond giving specifications, and discussing why it was never coated with a reflective aluminum surface, we identified the mirror as the backup which was, in fact, polished and figured correctly. In a section titled "A Flawed Mirror and an Ingenious Fix" we provided labels, images, and a human hair ("Hubble's mirror differed in shape by less than 1/50th the thickness of this human hair") to illustrate the magnitude of the flaw and how the flaw was corrected by the Corrective Optics Space Telescope Axial Replacement (COSTAR) and WFPC2. The rest of the panel dealt with the scientific harvest in sections titled, for instance, "Lengthening Our 'Cosmic Measuring Stick'," which introduced the Key Project in a way that we hoped would encourage public appreciation.

Figure E-5. After NASA offered the HST backup primary mirror to the astronomical community, finding that none of the interested parties were able to satisfy certain scientific and technical criteria to justify its use, they offered the mirror to the Smithsonian. Kodak Precision Optics personnel (left) donated expertise, time, and effort to bring the mirror from Perkin-Elmer and to prepare it for display (right) in the "Explore the Universe" gallery. (Smithsonian photographs by NASM staff.)

Originally we had hoped to place a backup HST secondary mirror on a pillar in a position appropriately juxtaposed to the primary, to give visitors a sense of the telescope's scale and how the components fit together. This plan was abandoned when the only known secondary mirror was sent on a long-term NASA travelling exhibit.

The Faint Object Spectrograph (FOS) originally on display in Space Hall was moved into "Explore the Universe" for its opening in September 2001. In "Explore" it became part of a cluster of instruments in a section on the detection of dark matter in the universe. The section covers instruments that played a role in searching for invisible matter, featuring Vera Rubin's spectrograph, an early Wolter lens from Riccardo Giacconi's sounding rocket payload that isolated Sco X-1, a one-fifth-scale model of Chandra and the FOS. Panels also portrayed "A Picture of a *Real* Gravitational Lens" and, as an introduction to the FOS, a section initially titled "The First Conclusive Evidence for a Black Hole" to describe how the FOS examined the accretion disk in M87. This title was later corrected, thanks to a suggestion from Harvey Tananbaum, to "Uncovering Evidence for a Supermassive Black Hole."

Hubble's Harvest

In anticipation of the 15th anniversary of the launch of the HST, the museum teamed up with the Space Telescope Science Institute to present a series of the best images from space gathered in by the HST's cameras. Ever

Figure E-6. When the WFPC2 was returned after the last servicing mission, it was displayed temporarily in NASM's Space Hall in front of the SDTV and a wall of Hubble images. (Smithsonian photograph by Mark Avino, NASM.)

since Jeff Hester and Paul Scowen of Arizona State University produced their image of the Eagle Nebula and its iconic "pillars of creation" in April 1995, NASA and the media have become keenly aware of the public impact the visuals offered. When we were planning "Explore," we appreciated this fact but decided not to present more than a few and to subscribe instead to Viewspace, which is managed by the Space Telescope Science Institute and is downloaded to a 50-inch monitor in "Explore" on a continuous basis.

The 15th anniversary, however, provided a new opportunity, and now we decided to display 10 of the best Hubble images from WFPC2 and the Advanced Camera for Surveys, portraying a progression of objects and fields from nearby planets to deep extragalactic space (Figure E-6). The Space Telescope Science Institute's Zoltan Levay provided large-scale prints, and with John Stoke we crafted a series of labels that were introduced by a panel that illustrated the three-color techniques for producing the images. Mounted in special low-reflection glass frames, these images span a wall in Space Hall extending some 40 feet from the Skylab installation to the window wall behind the display of the SDTV. Coincidentally, DeVorkin teamed up again with Robert Smith and Elizabeth Kessler to prepare a large pictorial volume published by National Geographic titled *Hubble: Imaging Space and Time*. In three editions thus far, it has sold more than 125,000 copies.[12]

With each of the servicing missions to the HST, not only has the telescope's performance been improved, but our chances for preserving parts of it, and its material legacy, have improved as well. The FOS was the first complete instrument acquired, returned from Hubble after the second servicing mission in 1997. Then, after the final servicing mission in May 2009, COSTAR and WFPC2 came available for loan for several months through the spring of 2010 and were displayed prominently in Space Hall in a new exhibition, "Moving beyond Earth." After that time, COSTAR remained at NASM as an accessioned object and WFPC2 was returned to NASA for exhibition and study. The WFPC2 was released by NASA in early 2014, and soon after, both it and COSTAR were displayed in front of the SDTV in Space Hall.

Final Thoughts

Just about all of the authors of the essays in this volume have greatly aided our display efforts as well as our historical research in past years. They appreciate that our historical interpretations, and our decisions about public display, are as transitory as the scientific knowledge base, which changes significantly over time. As new social and cultural perspectives emerge as new frameworks for interpretation, they will be applied to re-discuss aspects of history in contexts far different than those considered useful or allowable today. Future historians will be looking back on this time with a perspective about which we can only speculate.

It must be appreciated that since our display offerings at NASM reach a very large public, and one which is, frankly, not too deeply informed about the nature and practice of space research (though they are aware of its most spectacular products), we must take care not to present speculative or less than solidly confirmed perspectives. Unlike the print and electronic media, our offerings are also comparatively very expensive to produce, always requiring that the artifacts themselves are never compromised. Even in the case of altering the appearance of the SDTV, no changes were made that could not be reversed to bring the object back to the state in which it was acquired. These factors all demand that our exhibitry be highly constrained yet stimulating and provocative enough to be attractions our visitors will encounter and then hopefully ponder and maybe even absorb to the point where they dig deeper, into magazines and books and the electronic media, to further appreciate the legacy of the HST.

For these reasons, even though we are keenly aware of many issues and themes that could well be presented surrounding the HST that might stimulate even more public interest and consumption, we restrained our efforts to those themes we thought most important and most accessible during the life of the HST to show how it more than met its scientific promise. We hope and fully expect that someday, as Cold War fears continue to recede into history, we will be able to chronicle just how and why our nation decided that it could build an immensely expensive, complex, and challenging thing like the HST and how it decided that the HST would be 2.4 meters in diameter, employ CCDs, be launched by the Space Shuttle, and be serviced by astronauts. These issues have already been addressed in magazines and scholarly histories and might someday be addressable in school textbooks and in museum displays. Today we can only hope that these new themes and perspectives—raising questions about the relationship, and dependency, of the HST's technology upon our nation's satellite reconnaissance system or recounting the factors that led to the HST's alignment with NASA's Space Shuttle as the only means for deployment and servicing—will be addressed. Key to making this happen is ensuring the survival of the artifacts themselves, in the expectation that they will continue to stimulate the curiosity of future scholars as to why they exist and look the way they do.

The HST will be remembered not only for the science that was done but for its role in justifying the servicing mission concept that was, more or less, part of the justification for the Space Transportation System model, of which the Space Shuttle is the exemplar. Today we may well celebrate how it was saved, repaired, and upgraded in visits spanning some 16 years of time. But in the future, will this capability be seen as a positive and constructive step in the history of spaceflight or will it be seen as a reflection of the priorities of a space agency intent upon establishing human spaceflight as a permanent capability?

Our job is to preserve as much as possible about the era, in a manner whereby it will be useful. Of course, we also have to be mindful of the fact that the choices we make today about preservation will in some ways define the history that will be written in the future, if indeed future history will continue to depend upon evidence as the history of our times does. The best way we can hope to insure that this will happen will be to preserve the information, make it available, and do what we can to encourage people to listen and ask more questions.

Notes

1   On participant history among astronomers, see Marc Rothenberg, "History of Astronomy," in *Historical Writing on American Science*, Osiris, Second Series Volume 1 (1985), 130. See also Soraya de Chadarevian, "Review" of Joseph S. Fruton, "A Skeptical Bio-

chemist," *Isis* 87(3) (September 1996), 507–510; and DeVorkin, "History Is Too Important to Be Left to the Historians," in *"Organizations and Strategies in Astronomy"*, ed. A. Heck (New York: Kluwer/Springer, 2013), 417–440.

2   Curatorial file for "Space Telescope Model," Catalogue Number A19810041000, NASM Registrarial Files. "America's Space Truck—The Space Shuttle," label scripts, 2 March 1981 through March–April 1983, Script Files, Exhibits Division, NASM, courtesy of David Romanowski.

3   Robert W. Smith, The Space Telescope: A Study of NASA, Science, Technology and Politics. (Cambridge, UK: Cambridge University Press, 1989).

4   Joseph N. Tatarewicz to NASM Collections Committee, 24 October 1985, SDTV Files, Space History Division, NASM.

5   Joseph Tatarewicz, "Concept Script" 13 June 1988, SDTV files, Space History Division, NASM.

6   Joseph Tatarewicz, "Label Script HST/SDTV," 4 October 1986, p. 5, label 14, Space History Division, NASM.

7   See, for instance, Kevin C. Ruffner, "CORONA and the Intelligence Community: Declassification's Great Leap Forwards," https://www.cia.gov/library/center-for-the-study-of-intelligence/csi-publications/csi-studies/studies/96unclass/corona.htm (accessed 16 October 2012).

8   Gregg Herken, Art Molella, and Stanley Goldberg to Exhibits Committee, "Concept Proposal for 'Cold War' exhibit" 20 December 1993, NASM; Department of Space History to Exhibits Committee, 20 December 1993, NASM; "First Cut," n.d., author's Space Race Exhibit files, "Cold War Stuff," NASM.

9   Roger Launius commentary, excerpted in Valerie Neal to Paul Ceruzzi et al., 16 January 1996, author's Space Hall Exhibit files, NASM.

10   "Explore the Universe" label script, UN:624-L8-P8. NASM Exhibit Files.

11   Robert W. Smith and Joseph N. Tatarewicz, "Counting on Invention: Devices and Black Boxes in Very Big Science," in *Instruments*, Osiris, Second Series Volume 9 (1994), 101–123.

12   David H. DeVorkin and Robert W. Smith, *Hubble: Imaging Space and Time*, with contributions by E. Kessler, (Washington, D.C.: National Geographic, 2004, 2008, 2011).

Appendix: The Decision to Cancel the Hubble Space Telescope Servicing Mission 4 (and Its Reversal)

Steven J. Dick

Introduction

On 16 January 2004 NASA Administrator Sean O'Keefe announced his decision to cancel the Hubble Space Telescope (HST) Servicing Mission (SM4) by the Space Shuttle.[1] The SM4 was to have inserted two new instruments, the Wide Field Camera 3 and the Cosmic Origins Spectrograph, at the same time replacing the batteries and gyroscopes, extending the HST's lifetime to 2010. The decision resulted in a strong reaction among some members of Congress, the HST science community, and the general public because it would likely leave the telescope inoperable by 2007, years before its full lifetime and well before the James Webb Space Telescope (JWST) would be launched. What follows is a history of that decision and its aftermath.

Background

After a long history of concept, design, and construction stretching back to 1965, the HST was launched 24 April 1990.[2] Scheduled for launch in late 1986, it had been delayed by the Space Shuttle *Challenger* disaster in January of that year. Although there had been other successful telescopes in space, notably the Orbiting Astronomical Observatories 2 and 3 (Copernicus) in the 1960s and 1970s, the HST, with its 2.4-meter mirror and more than $1.3 billion price tag, was in a different league. Disappointment was therefore acute, to put it mildly, when it was discovered shortly after launch that spherical aberration in the mirror made the HST images blurry, greatly limiting its scientific capacity. The press had a field day ridiculing NASA and its engineers, a situation that was not helped when the subsequent investigation discovered that faulty testing of the mirror had been the culprit.

Hubble Servicing Missions

Ever since it became clear that it would be launched with the Space Shuttle rather than a Titan III rocket, Hubble's fortunes had been bound up with human spaceflight. The good news in the HST's bleak situation after launch was that it had been designed to be serviced. The triumph was all the greater when, in December 1993, the first Hubble servicing mission (SM1) succeeded in placing corrective optics into the telescope, rendering its new images perfect. It was not only vindication for the HST but also for the concept of human servicing. So high were the stakes, some called it a "save NASA" mission.[3] Over the next decade three more servicing missions followed. The SM2, carried out with the shuttle *Discovery* during Space Transportation System (STS)-82 in 1997, was the highest Space Shuttle flight, reaching an altitude of some 386 miles (~621 km). It was this mission that President Bush indirectly referred to in his 2004 space exploration speech, when he cited 386 miles as the furthest humans had been from Earth since the last Apollo mission in 1972, a quarter century earlier. On this mission the NASA Goddard High Resolution Spectrometer and Faint Object Spectrograph were replaced by the Space Telescope Imaging Spectrograph (STIS) and Near Infrared Camera and Multi-Object Spectrometer.

What was to be the third HST servicing mission was broken into two missions, SM3A and SM3B, later causing some confusion among the media and public with regard to the number of servicing missions. The SM3A, carried out in late 1999 with the shuttle *Discovery* during STS-103, took place under urgent conditions and was moved up in the schedule to accomplish that part of the original SM3 mission that needed to be done immediately. The telescope itself was in safe mode, its gyros having failed, and the servicing mission had to be accomplished before the end of the year because of millennium (Y2K) software fears. The crew successfully installed new gyroscopes and scientific instruments, and the telescope was redeployed on Christmas day. The SM3B, the fourth HST servicing mission, was carried out in March 2002 during the *Columbia* STS-109 flight. It installed a new digital camera, a cooling system for the infrared camera, new solar arrays, and a new power control unit. The latter was a particular triumph, since it went beyond the normal servicing requirements. Payload Commander John Grunsfeld recalled, "Nobody believed we could necessarily do that; this is a big switch box, lots of connectors, all the power runs through it, and there was a problem with it that would, gone unchecked, have terminated Hubble's life early, probably in the 2005 to 2008 timeframe. And we took that issue all the way to the administrator, at that time Dan Goldin, and said this is a tough one; if we try this and it doesn't work we lose Hubble; if we don't try it we'll probably lose

Hubble. And it's well beyond the limits of any kind of EVA [extravehicular activity] that's ever been done, harder, longer, and it involves significant risk to the telescope. And Dan Goldin looked at me straight in the eyes and said, 'Well John, do you think we can do it'?" Grunsfeld answered in the affirmative, and though he characterized it as "the most challenging space walking activity we've ever done in the space program," it proved very successful.[4] As it turned out, Grunsfeld was the last person to touch the HST.

Sean O'Keefe

There was another novelty to the SM3B mission. After a record 10 years as NASA administrator, Dan Goldin had left the agency the previous November. The STS-109, with its HST servicing mission, was the first opportunity for his successor, Sean O'Keefe (Figure A-1), to witness a shuttle launch. O'Keefe had joined the administration of George W. Bush on inauguration day and served as deputy director of the Office of Management and Budget (OMB) until his appointment as NASA administrator on 21 December 2001. It was his fourth presidential appointment, having also served as comptroller and chief financial officer of the Department of Defense (1989) and Secretary of the Navy (1992). He had also served for eight years on the U.S. Senate Appropriations Committee staff and as the Louis A. Bantle

Figure A-1. Sean O'Keefe, NASA administrator between 2001 and 2005. (NASA image GPN-2003-00090; http://grin.hq.nasa.gov/ABSTRACTS/GPN-2003-00090.html.)

Professor of Business and Government Policy, an endowed chair at Syracuse University's Maxwell School of Citizenship and Public Affairs. With this background O'Keefe was in a strong position to bring NASA's budget under control, in particular cost overruns on the International Space Station (ISS), which had subjected NASA to severe congressional criticism during the Goldin years. And with the STS-109 as his first shuttle launch, O'Keefe was well aware of the importance of the HST servicing missions from the beginning of his tenure.[5]

The Columbia Accident Investigation Board (CAIB) and the Stafford–Covey Return-to-Flight Task Group

The next servicing mission, designated SM4, was to have been carried out in November, 2004, but disastrous events intervened on 1 February 2003 with the catastrophic loss of *Columbia* and its crew. Administrator O'Keefe was at

Kennedy Space Center waiting for the landing, which never came. Shortly after the planned landing time of 9:16 a.m. he declared a shuttle contingency, and the Action Plan for Space Flight Operations was implemented. Within hours of the accident he appointed an investigation board, named the following day the Columbia Accident Investigation Board (CAIB; Figure A-2), to be chaired by Admiral Harold W. Gehman Jr. Gehman was a retired four-star admiral who had served as the NATO supreme allied commander, Atlantic; commander in chief of the U.S. Joint Forces Command, and vice chief of naval operations for the U.S. Navy. He had co-chaired the Department of Defense review of the attack on the U.S.S. *Cole*. The CAIB was charged with investigating the facts and probable causes of the accident and with recommending "preventative and other appropriate actions to preclude the recurrence of a similar mishap."[5] After a seven-month investigation the board issued its report 26 August 2003. Among the many recommendations was the following: "For non-station missions, develop a comprehensive autonomous (independent of Station) inspection and repair capability to cover the widest possible range of damage scenarios."[6] Although the HST was not mentioned by name, the only post-*Columbia* missions that would not fly to ISS were the servicing missions to the HST. As with all of the recommendations, O'Keefe was to take this one very seriously.

Meanwhile, on 13 June 2003 O'Keefe established the Return to Flight Task Group, whose charge was to implement the recommendations of the CAIB report. Chaired by two veteran astronauts, Thomas P. Stafford and Richard O. Covey, the group would undertake numerous fact-finding visits, public meetings, and media teleconferences. Most importantly, it produced "NASA's Implementation Plan for Return to Flight and Beyond," a "living document" first released on 8 September, followed by interim reports in

Figure A-2. The Columbia Accident Investigation Board convenes for a third public hearing at Cape Canaveral, Florida, as part of some seven months of deliberations. The board heard testimony, made inspection visits, and analyzed exploratory tests of Shuttle materials. Navy Admiral Harold W. 'Hal' Gehman Jr. chaired the board. From left to right: Steven B. Wallace, Scott Hubbard, Dr. John Logsdon, Rear Admiral Stephen Turcotte, Gehman, General Duane Deal, Dr. Douglas Osheroff, and Major General Kenneth W. Hess. Board members not present in this session were Major General John Barry, Dr. James N. Hallock, Roger Tetrault, Dr. Sheila Widnall, and Dr. Sally Ride. (NASA image GPN-2003-00079; http://grin.hq.nasa.gov/ABSTRACTS/GPN-2003-00079.html.)

January and May 2004.[7] The recommendations of the CAIB report were the benchmarks against which NASA's progress would be monitored on a point-by-point basis in return to flight (RTF) meetings held by the Stafford–Covey group and at headquarters. Those meetings would play a crucial role in the HST SM4 decision.

Other Studies Related to Post-*Columbia* HST

In addition to the recommendations of the CAIB report, several studies were chartered to analyze the future of the HST, and these naturally had to take into account the impact of the *Columbia* accident. The first arose from NASA congressional appropriations language in February 2003: "The conferees direct NASA to carry out an in-depth study of an additional servicing mission (SM5) in the 2007 timeframe that would study operating HST until the Webb Telescope becomes operational. The study should address the costs of an additional servicing mission and the potential scientific benefits." This "HST Post SM4 Scientific Review Panel," as its name implied, was to deal with longer term issues. Also termed the Black Commission after its chair, David Black, in April 2003 the commission assumed that SM4 would be conducted in the 2004–2005 time frame. It concluded that HST would continue to provide high-quality science even beyond the time of a proposed SM5 but foresaw budgetary and technical problems with a servicing mission in the 2007 time frame.[8]

In June the Office of Space Science, realizing that "it is a necessary task to consider exactly how and when to terminate the operation of this successful scientific experiment," chartered the HST–JWST Transition Plan Review Panel, chaired by John Bahcall, to evaluate the scientific impact of the current NASA plan for ending HST operations and beginning the JWST operations. That plan called for the end of the HST operations in 2010 and the launch of JWST in late 2011. In August the panel provided three options in priority order. (1) Two additional shuttle servicing missions conducted, SM4 in about 2005 and SM5 in about 2010, in order to maximize the scientific productivity of the HST. The extended HST science program resulting from SM5 would occur only if the HST science was successful in a peer-reviewed competition with other new space astrophysics proposals. (2) One shuttle servicing mission, SM4, before the end of 2006, which would include replacement of the HST gyros and installing improved instruments. In this scenario, the HST could be de-orbited, after science operations are no longer possible, by a propulsion device installed on the HST during SM4 or by an autonomous robotic system. (3) If no shuttle servicing missions were available, a robotic mission to install a propulsion module to bring down the HST in a controlled descent when science is no longer possible.[9] The conclusions of the report were endorsed by the American Astronomical Society, which strongly urged that whatever support was needed for SM4 should be found, consistent with CAIB recommendations.

The Bahcall panel reported its conclusions in mid-August. Less than two weeks later the CAIB issued its report, and it was this report and the RTF issues that were destined to have the greatest impact on the final decision to cancel SM4. That decision, which would eventually be made, was similar to the third and last priority in the Bahcall report.

The Decision

In the minds of several of the key players in the decision, the first thought that SM4 might be canceled dated to the *Columbia* disaster itself. As Ed Weiler, associate administrator for Space Science put it, "I got a first inkling that the servicing program in general was in trouble on 1 February 2003 when I turned on CNN [Cable News Network] in the early morning and saw what was an unmistakable signature of a spacecraft breaking up in front of my eyes...and I knew at that point that if it was what I thought it was, which was the destruction of the shuttle, that this would portend poorly for future shuttle flights to orbits like Hubble's orbit. I was certainly worried about it." Along with concern for the astronauts, it was natural for Weiler to think about the ramifications for the HST, which came under his Office of Space Science. The same thought must have been in the minds of the other HST managers also, since there was no way to service the HST without the shuttle. Everyone knew the *Challenger* accident had caused a long delay in the RTF. Fortunately, at the time of the *Columbia* accident, the HST had been serviced less than a year earlier; still its batteries and gyros would inexorably wear out, and there was no doubt of the importance of timeliness for another servicing mission. A post-*Columbia* RTF date would depend on the course of the investigation and the cause of the accident, and in this respect the recommendations of the CAIB report would assume utmost significance.

The Role of the CAIB Report and Return-to-Flight Meetings

Administrator Sean O'Keefe recalled that for him the decision process for SM4 began in a serious way...

> on August 26, 2003, when the Columbia Accident
> Investigation Board released its report. So we started
> looking through all those challenges, consistent
> with all the return to flight [RTF] activities we were
> engaged in as early as March–April 2003, when the

formal kind of framework got kicked off lining up an RTF process. It wasn't directly informed by all the recommendations, findings and observations until the 26th of August. Then thereafter each step along the way we were formulating a regular assessment that began in September 2003 of what it would take in order to implement those recommendations to RTF. And as each mounting month went by at every update of the return to flight document...every one of those reveals it is harder and harder and harder to accomplish every one of those recommendations to achieve that objective. So I think by the late fall, early winter, it was pretty apparent that our likelihood of accomplishing all those objectives in time to mount a servicing mission that would be in compliance with all those recommendations was becoming more and more remote.[10]

Bill Readdy, the associate administrator for Space Flight who had himself flown three shuttle missions between 1992 and 1996, also pinpointed the CAIB report as the event that triggered serious discussion about the Hubble mission. Although the CAIB investigators "were not saying that we couldn't fly it if we developed stand-alone autonomous inspection and repair capability...the bit was pretty much set in my mind that this was going to be a very, very high bar set to ever go do a Hubble servicing mission."

More broadly, Readdy was struck by the CAIB's finding "that NASA's not a learning organization. That NASA failed to completely follow up on the *Challenger* recommendations. I was left with a clear impression that, yeah, we could proceed at risk to go off and do another Hubble servicing mission, but that would also be conclusive proof that NASA hadn't learned anything from *Columbia*...or from *Challenger* by implication."[11]

The return-to-flight meetings made it clear that there were numerous obstacles to a quick resumption of shuttle flights. The prime objective was not speed but safety, and that meant at a minimum satisfying all of the CAIB recommendations.

Assessment of SM4 Options

In the wake of the Bahcall and CAIB reports, in the midst of the much broader and still moving target of a return to flight date for the shuttle, the assessment of an SM4 decision continued, both within the Office of Space

Science and at higher levels. By early November it was still more a question of when, not if, such a mission would occur. On 7 November Weiler presented O'Keefe with the advantages and disadvantages of dates for a servicing mission ranging from 2005 to 2007. They talked about how long the gyros would last, and Weiler recommended one more servicing mission but not another one (SM5) beyond that. "I pointed out that if you know you are only going to have one more gas stop and you want to go as many miles as you can, do you fill your gas tank up when it is half full or do you wait until you're on fumes? That is the argument that said, if you wanted to wait until you were on fumes you would probably go to maybe 2007, but that was pushing the envelope, so we centered on the optimum time...of the U.S. Core Complete for [the International Space] Station, which would be around June 2006, and that is where the June of 2006 came from. If we planned a servicing mission June of 2006, it wouldn't impact finishing off the station, Core Complete U.S., making sure we got it restocked with water and food." According to that brief, the latest useful SM4 mission would have been 2008, but a reliable restart of the spacecraft would have been in doubt by then. And there was a final option listed: "no SM4." A backup slide—to be used only if necessary—gave the story of the HST without SM4: experience showed that ways would be found to extend its life; thousands of archival images existed that astronomers could still study; and savings could benefit other programs. But the slide was either not used or in any case did not carry the day. "We gave that presentation to him and we said the way we would dispose of Hubble is we wouldn't plan a shuttle anymore because obviously that would be crazy. We would build a robotic thing to grab it and take it to the Pacific and he approved that. I left that meeting ... feeling like we were on the road to an SM4."[12]

The NASA Fiscal Year 2005 Budget

As the return to flight meetings were proceeding, and the HST managers were assessing their options, the fiscal year 2005 (FY05) NASA budget was being prepared. Per the usual procedure NASA submitted its budget to the OMB in September, and OMB gave NASA its "pass back" with revised numbers in November. Thanksgiving weekend saw NASA Comptroller Steve Isakowitz, O'Keefe, and others finalizing the budget to go back to the White House to get the president's approval before it went on to Congress. February first was the traditional day when the final budget was rolled out; ironically, it would be the first anniversary of the *Columbia* accident.

The SM4 had budget implications; if it were going to be in the 2005 budget, "offsets" needed to be found in other areas of space science, something that the Office of Space Science was perfectly willing to do. But, Isakowitz recalled,

> the problem was that if you went strictly by what came out of the CAIB recommendations in terms of the ability to inspect and repair and safe haven, we had no known way to do it. So we can go ahead and budget for a date, but then the question becomes when would we actually know that we could fly it? ...As we began to ask questions like that, even then it became clear that we wouldn't know maybe until the last minute whether or not we could actually do such a mission. Yet in the meantime, we are going to have to spend lots of money and keep it all going.[13]

By this time, then, the RTF implications for an HST mission were coming to the fore. Isakowitz left the Thanksgiving weekend meeting with O'Keefe with a tentative decision that the SM4 would not be in the budget. It was, he said, one of "a million other decisions" on the list. Asked if the SM4 was a budget issue, Isakowitz said "No, the only reason I would say it is tied to the budget was the budget helped to dictate the timing of when we were going to make a decision." Elaborating further, he noted,

> That is what the budget process does. When you have issues, even if it has nothing to do with the budget...the budget process forces people to make decisions.... The budget dictated the schedule as to when the decisions were going to be made. For those who...still argue that this was a budget decision, we cut the Hubble to pay for the vision, that is just simply not true. We would have found the money to do the Hubble.[14]

As O'Keefe put it, referring to the Thanksgiving weekend meeting,

> The choice was you either had to put the resources in to continue planning for that mission through FY05 or not. And it finally got down to the point where the act of leaving it as it was would have signaled improperly that we had planned to do a mission that I had come to the conclusion that I didn't think we were likely to be able to do...could have been delayed...but in the end ultimately it would have had to be manifested in that way to make a decision. So it was not a question of whether you put how much in, it was a question of

whether you put anything in...I realized at that stage of the game that if I did not make that decision at that time it would be potentially another year that we would maintain the fiction that we could do this mission.

O'Keefe called that meeting "a prompting event," a way of forcing him to make a decision, but added that in the end, the [decision] was based on the unlikelihood of meeting the CAIB recommendations before the predicted turnoff of the HST.[15]

The Decision Made

Asked the date when the final decision to cancel the SM4 was made O'Keefe said it "probably converged around the early part of December," after the RTF meetings showed more and more clearly that it could not be done in time to save the HST. It was at a crucial 2 December meeting of the Executive Committee, where Isakowitz briefed NASA associate administrators on the 2005 budget submission, when it first became clear at the associate administrator level that the SM4 mission was not in the budget. "That was the first time I saw that SM4 was cancelled and that was the first time anybody in that room other than Steve [Isakowitz], I guess, and Sean, knew that SM4 was cancelled, so I had to react in real time," Weiler recalled. Asked if he felt he was not consulted Weiler replied,

No, because I could have stood up at that meeting. Nothing was published at that point in time. I could have said I object. I think it is safe. I think the science is worth it, but that would be disingenuous of me because I don't know if it is safe or not. I'm not a safety engineer. I think it is very important for people to recognize their own limitations. I'm going to be an enemy of the scientific community because of this. I could get up there and be on my high pulpit and say damn with safety, we have to go fix the Hubble because it is the greatest scientific thing since sliced bread. I could say that but that is the easy way out. That is the easy way out, hide behind the science.[16]

Still, it had to be a difficult decision for Weiler, who had been associated with the HST project for 25 years:

Anybody who says I take this lightly is missing the point. I am taking it rationally not lightly. I cannot stand up and say that the science justifies additional risk. I don't know how to quantify science in those terms. Human life is too valuable...I wouldn't want to have to explain to a four-year-old boy why he will never see his dad again, or his mom. That has to be the position Sean was in. That is a serious position to be in. It is a lot different than sitting in an ivory tower university making pronouncements about how valuable the science is. That is as blunt as I get.[17]

O'Keefe later confirmed this aspect of his thinking, when responding to what some called the "withering" criticism of the SM4 cancellation: "Let me offer my view of 'withering'," he said.

Withering is the feeling you get when you are standing at a runway with the dawning realization that the shuttle everyone is waiting for isn't going to land. Withering is when you have to explain to wives, husbands, parents, brothers, sisters, and children that their loved ones aren't coming home alive. Withering is attending funerals, memorial services, and ceremonies over 16 months in number too many to count any more, yet having every single one of these events feel like the weight of that responsibility will never be relieved. Withering is the knowledge that we contributed to the *Columbia* disaster because we weren't diligent.[18]

In O'Keefe's estimation, every further RTF meeting confirmed the wisdom of the SM4 cancellation decision. In particular the RTF meeting at Johnson Space Center on 12 December, following the Stafford–Covey Task Group fact-finding visit there the previous three days, confirmed that the CAIB recommendations were not likely to be met by the hoped-for September–October timeframe.[19] At about the same time the Space Flight Leadership Council (the spaceflight community) concluded that the RTF would not occur in September–October of 2004 but would likely slip to March–April of 2005. As O'Keefe recalled, "All those events were converging in that few weeks span of time, and looking more and more and more apparent that the likelihood of return to flight in a timely manner was remote and therefore even more so remote that you'd be able to mount a servicing mission unique to Hubble." On 19 December, during a brief on the still unannounced new exploration strategy for NASA, O'Keefe informed the president that the HST mission was not going to happen. The president agreed that compliance with the CAIB recommendations was paramount.[20]

Plans for Announcing the SM4 Cancellation

Planning for the SM4 cancellation announcement fell to NASA's chief scientist, John Grunsfeld. Curiously, as January began Grunsfeld had little idea what was about to happen to the HST. An astronaut who had participated in the last two HST servicing missions (SM3A and SM3B in 1999 and 2002), prior to becoming NASA's chief scientist in September 2003 Grunsfeld had been leading the activity for the SM4 at Johnson Space Center, home of the astronauts. In the summer he had testified before the Bahcall group, saying there was astronaut consensus that the SM4 was one of the missions "worth risking our lives for...really important for humans to do...the marriage of human spaceflight and robotic science spaceflight."[21] By contrast the astronaut office was not on board for risking lives for any mission to bring the HST back for the National Air and Space Museum.

Throughout the fall, in his position as chief scientist, Grunsfeld discussed with Anne Kinney, head of the Astronomy and Physics Division of the Office of Space Science, the details of carrying out the SM4. Neither had any inkling it might be cancelled except for the general rule that no mission was secure until it actually flew. At the same time he had urged the community to concentrate on the SM4 rather than worrying so much about the SM5. Although Grunsfeld had gotten a faint signal from Isakowitz during the OMB budget pass back around 28 November that the SM4 might not be in the budget, only on 7 January was he informed in an abrupt way. The previous day Grunsfeld was at the winter meeting of the American Astronomical Society in Atlanta when he got a Blackberry message inviting him to a senior staff meeting the following day to discuss the HST servicing mission timing.

Grunsfeld immediately flew back to Washington from Atlanta. He had assumed the meeting was to discuss the timing for the SM4 in the flight manifest, but when he walked into the meeting it was clear that decision to cancel the servicing mission had already been made and the discussion was how to roll out the decision to the public. Grunsfeld was stunned; he "literally felt like somebody hit me in the head with a two-by-four."[22] Moreover, because Ed Weiler's Office of Space Science was about to land two rovers on Mars, Grunsfeld was given the unhappy task of coming up with a plan of how to roll the decision out to the public. Grunsfeld consulted with some of his mentors, including John Bahcall, as to whether he should even stay with NASA in the wake of such a decision on which he had not been consulted.

He decided that matters might be worse for the HST if he left. Thus, over the course of several senior staff meetings he laid out a plan that would be rolled out on 28 January, a few days prior to release of the president's budget, at a press event that O'Keefe would lead. Prior to that, the HST principals would be informed in an orderly way.

The Role of Probabilistic Risk Assessments versus Intuition

Meanwhile Grunsfeld went to his fellow astronaut Bill Readdy (Figures A-3, A-4), the associate administrator for Space Flight, the office in charge of shuttle flights, looking for a probabilistic risk assessment (PRA) that might document the risk. A PRA is a comprehensive, structured, and logical analysis method aimed at identifying and assessing risks in complex technological systems for the purpose of cost effectively improving their safety and performance. It was a computer model tailored for each technological case, used for years in the nuclear industry, and since 1995 at NASA in relation to the shuttle. As Brian O'Connor, chief safety and mission assurance officer, put it, a PRA...

> incorporates all the best technical know-how of your system, how it's hooked up, inter-relationships between subsystems. For example in the model if you fail an electrical circuit, just take it out [—] you can do this in these PRA models [—] you can fail things. Then it can have an effect on your thermal system and your navigation system, it takes away a leg of redundancy from your cooling loops and all the kinds of things because it's just a big software model of your system. And the way the probabilistic risk assessment works is that it takes all of the best notions of your engineering and your safety and reliability community on failures and what chances they have of failing, and it factors in all these accident scenarios that could happen.[23]

A PRA was not comprehensive in every detail, however; while the chance of loss of a thermal protection system was in the shuttle model, the risk due to insulating foam from the external tank hitting the shuttle was not.

In this case, no such analysis existed. According to O'Connor the shuttle PRA was going through a peer review and was not useable for testing this scenario. Even if it had been possible to compare the risks of an HST rescue mission to a space station mission, he noted, a PRA was only one piece of the puzzle:

Figure A-3. Astronaut Eileen M. Collins, STS-114 commander, and William F. Readdy, associate administrator for Space Flight, NASA Headquarters, converse while waiting for the STS-114 crew return ceremonies at Ellington Field near Johnson Space Center (JSC) in 2005. (NASA image JSC2005-E-33437; http://spaceflight.nasa.gov/gallery/images/shuttle/sts-114/html/jsc2005e33437.html.)

> Far be it from me to ever suggest that anybody would ever make a decision like this just based on risk trade from a PRA, because I know that a PRA is limited as a model, it only looks at certain things. It doesn't look at some of those secondary things like the distraction factor of putting a different kind of mission, and all the planning that goes with it, in the middle of your return to flight activities to the station. You now have tasked your people to go worry about other things like how you do a shuttle to shuttle safe haven rescue, which you wouldn't worry about on station. On station you know how to hook the shuttle up to the space station to get the people out, but we've never thought much about how you would go up there and bring another shuttle up to a crippled shuttle and get the people out of one vehicle into the other, so a lot of work would have to be done there and there's risk inherent in that. It's not even in this model.[24]

Grunsfeld came to understand that O'Keefe's decision was an intuitive call: he had synthesized the RTF data and concluded that it was too hard. Asked whether his decision was intuitive, O'Keefe answered:

Absolutely, no question. But rather than calling it "intuitive grounds," I would say "intuitive" in the sense of confidence level and attaining the objectives of the Accident Investigation Board recommendations as a forecast in time. That part is intuitive; you can't analytically demonstrate whether you will or you won't.... You kind of look at what the trend-line looks like at any number of things...so it is by nature more of an intuitive circumstance of where you see the trend going...it is driven by the analysis and the data and the information and the current status of our capacity to do things technically.[25]

The New Space Exploration Vision

Meanwhile, events were occurring that would have a profound effect on NASA's future. In the wake of *Columbia*, and especially after the CAIB report in August, the White House was planning a new exploration vision for NASA. In the summer of 2003 a White House interagency team began meeting to consider the options. Among the options was phasing out the shuttle by 2010, something that could obviously impact the HST.[26] All of this was coming to a head in January 2004 at the same time that the senior staff meeting was being held to decide how to roll out the SM4 decision. On 13 January the Leadership Council, including NASA Headquarters leaders and center directors, were briefed on the president's space exploration vision. Isakowitz presented the details of the budget implications, and although he did not mention the HST, "it came up. There was some discussion at that point and that was the meeting at which people talked it through because that was where some people were hearing it for the first time."[27]

On 14 January President Bush came to NASA Headquarters to announce the new space vision (Figure A-5). It included retiring the shuttle by 2010, abandoning the space station around 2016, and sending humans to the Moon by 2020 and to Mars by some unspecified date.[28] The same day a staffer from the White House went to Capitol Hill to brief staffers on the vision. When he talked about the shuttle flying until only the end of the decade, someone asked about the implications for the HST. The staffer said this meant cancellation of SM4, and on 15 January an article in the *Washington Post* mentioned this fact in passing, obviously having been leaked. In speculating on the possible implications of the president's vision, Kathy Sawyer wrote, "There may also be slowed growth in the NASA space science budget, sources said, and a 'refocusing' of activities within the agency to support the central theme of returning to the moon. There will be no further servicing missions to the Hubble Space Telescope. Though there is rampant speculation about closing NASA facilities and axing programs, there were few specifics."[29]

This "accidental release" unleashed a variety of charges—that HST was being sacrificed for the new vision, that it was a victim of budget cuts, and so on. Grunsfeld admitted that it looked bad for NASA. "I think to the press that looks suspicious, Friday night late calls as if we were trying to pull a fast one; and from there it has been an uphill struggle."[30] He was right about that.

Decision Announcement

Grunsfeld's carefully crafted plan for the announcement was shattered. Worse than that, even some of the principals involved in the HST servicing missions found out the hard way. Michael Moore, the program executive for the HST at NASA Headquarters, heard it from his boss, Anne Kinney, on 15 January, the day after the president's speech and the same day as the *Washington Post* article. Jennifer Wiseman, the program scientist for the HST at NASA Headquarters, found out from Michael Moore via a telephone call very early that same morning of 15 January.[31] Moreover, those who operated the HST at the Space Telescope Science Institute in Baltimore, and those who planned the servicing missions at Goddard Space Flight Center (GSFC), were totally in the dark. On Friday morning, 16 January, O'Keefe, Weiler, and Grunsfeld made the short trip to GSFC, where they broke the news to the "Hubble team." Among those present in the audience were Steven Beckwith, director of the Space Telescope Science Institute, and Frank Cepollina, a 40-year veteran of NASA who had been in charge of all the HST servicing missions. Also present from NASA Headquarters were Anne Kinney, Eric Smith, and Jennifer Wiseman. Trying to make the best of an admittedly bad situation, O'Keefe spoke for about 45 minutes, without notes, saying the decision was his alone. He asked the Hubble team to come up with creative ways to extend Hubble's life by increasing efficiencies in the batteries or gyros or both. The administrator was followed by John Grunsfeld (Figure A-4) and Ed Weiler, who endorsed the decision. There were questions and answers, and the mood was somber. As Wiseman recalled, "It had the very same sort of funeralesque type of atmosphere where people were somber and yet they tried to comfort each other with small statements of comfort."[32]

The timing was unfortunate in terms of the president's space exploration vision announced two days before. As Isakowitz recalled, "It was a tough decision and in an ideal world actually, it would have been great if we could have deferred it, because the fact is that the Hubble decision really had...no specific link to the vision itself, but it was clear that if we were going to take a decision that said not to do it, it would cast a shadow on the vision." But, like the budget, Isakowitz insisted that the HST decision and the new space exploration vision were unrelated. Bill Readdy agreed: "They were totally decoupled, they really were."[33]

The Reaction

The immediate reaction to O'Keefe's hurried 16 January announcement at GSFC of the cancellation of SM4 was swift and overwhelmingly negative. Perhaps most surprising was the reaction from the media and the public, which had pummeled Hubble because of its problems 14 years earlier but had now grown accustomed to the awe-inspiring pictures beamed down

Figure A-4. John Grunsfeld, photographed here during the STS-109 mission, had worked tirelessly to assure the continuation of the HST for as long as possible. (NASA image STS109-E-5419; http://spaceflight.nasa.gov/gallery/images/shuttle/sts-109/html/s109e5419.html.)

regularly from the orbiting telescope. Officials from NASA and the Space Telescope Science Institute received thousands of e-mails, some offering money. "The overwhelming amount of general public comment we've gotten is just sort of shock," Bruce Margon, associate director of the Space Telescope Science Institute was quoted as saying in the *Washington Post*. "If it's working," people ask, "how can you possibly shut it off? I don't have an answer to that," Margon said. He announced a public web site to accept public suggestions for the Hubble.[34]

Some in the media immediately linked the HST decision to the president's new vision for space exploration. In its cover story announcing the new vision, *Time* magazine wrote, "The budgetary shake-up has already claimed a victim. The Hubble Space Telescope had been scheduled for a maintenance visit next year by space-shuttle astronauts. Now there is no money for the mission, and after 2010 there will be no shuttle anyway. One of NASA's greatest success stories, Hubble will probably wink out sometime in 2007." [35] Like Readdy and Isakowitz, O'Keefe was adamant that the two were unrelated.

No! I was reading in *The Washington Post* an article that made very scant reference to it...it wasn't even an essential piece of it, it was just kind of a throwaway line. And I thought 'well hell, that's it, this accelerates the whole thing.' We ought to go out and describe this as quickly as possible...the last thing I wanted [was for] anybody to find out about this is from reading something in the paper....So I called [Senator Barbara] Mikulski, made arrangements to go up to Goddard, talk to the Hubble team, that's what accelerated everything at that point because originally we were thinking about trying to organize something that would be in sequence sometime the last ten days or so of January. All that went up in flames on the basis of this. There was no relationship, association or decision about this as it related to the president's vision statement or anything else. No, no linkage at all.[36]

O'Keefe was not surprised about the public reaction, but he was surprised by "the depth of personal animus described by the leading advocates of all this, it wasn't a professional issue to them, it's personal." In particular he was surprised by the animus in the "save the Hubble" petition on the internet.[37]

Perhaps less surprising was the reaction from the scientists most directly involved in Hubble. In a statement prepared for distribution to members of the American Astronomical Society on 28 January, Space Telescope Science Institute Director Steven Beckwith wrote, "the decision to end Hubble is a blow to astronomy and to NASA's efforts to engage a larger public in its mission of exploration and discovery. Never in the history of astronomy has society shut down its most powerful optical observatory before a successor was ready." The reaction at NASA Headquarters was understandably more muted, even from those not involved in the decision. Anne Kinney wrote that the Astronomy and Physics Division of the Office of Space Science was greatly saddened, but "fully supports the administrator's decision, a decision based on issues related to risk."[38]

Another argument, both among scientists and in the media, was that the HST needed to stay operational until its successor, the James Webb Space Telescope (JWST), was launched. Weiler had a ready answer to that:

[The] JWST was never sold as a replacement for the Hubble. It is a different kind of science. It is different wavelength color. It is a different community and these people who are out there again in the ivory tower saying we have to wait until the replacement for Hubble, they have a long wait because JWST isn't a replacement of Hubble. More importantly, where in the constitution does it say that optical astronomers always have to have a telescope in space? X-ray astronomers haven't got that in the constitution. Gamma ray astronomers don't have that. Infrared astronomers don't have that. It is ironic that optical astronomers, unlike those other areas I just mentioned, can do their astronomy from the ground too. The others can't.[39]

It was a remarkable statement from one who was not only an optical astronomer but who had spent a good part of his career on the HST.

The link to the president's space vision would not go away, even in the scientific literature. The respected professional magazine *Physics Today* headlined its story with "Hubble Sacrificed in Wake of President Bush's New Space Vision." Beckwith was again quoted as saying the SM4 cancellation "was a complete shock and devastating for everyone." John Bahcall, who had headed the panel reviewing the HST's lifetime options, including the option that SM4 might not occur, noted, "All the astronomers I talked to assumed SM4 would occur....I certainly assumed that it would. I think it is regrettable that no research scientist was involved in the decision to terminate the life of the most famous research telescope of the past 100 years."[40] *Science*, the professional magazine of the American Association for the Advancement of Science, surveyed the potential "collateral damage" from the new space vision and characterized the HST as one of the "unpleasant side effects." The new vision "would end the Hubble Space Telescope's brilliant run and could jeopardize the future of any activity that doesn't' directly serve the exploration effort."[41]

Congressional Reaction

Undoubtedly more disturbing to NASA managers was the reaction from Congress, whence NASA receives its funding. On 21 January Senator Barbara Mikulski, a Democrat from Maryland whose district included the Space Telescope Science Institute and the ranking minority member of the Appropriations Committee's Subcommittee on Veteran Affairs, Housing and Urban Development, and Independent Agencies, fired off a letter to O'Keefe asking him to reconsider the decision and appoint an independent review panel. "I was shocked and surprised by your recent decision to terminate the next scheduled servicing mission of the Hubble Space Telescope (HST)," Mikulski wrote. On 26 January O'Keefe replied, explaining his rationale in detail. "The

Figure A-5. The administrator of NASA, Sean O'Keefe (right), with President Bush after he announces the new vision for space exploration, 14 January 2004. (NASA image; credit: NASA/ Bill Ingalls; http://www.nasa.gov/exploration/about/how_we_got_here.html.)

decision had to balance the world-class science that HST has produced, and would continue to produce, against the risks to the shuttle and its crew. In the end, the determining factor was the recommendations of the Columbia Accident Investigation Board report for developing on-orbit inspection, repair, and contingency rescue requirements for every shuttle flight. As such, my decision was not made with regard to budget considerations, nor was it based on any question as to the significance of the science return of the HST."[42]

O'Keefe specifically made the following points:

1. Because the SM4 mission would have been the sole remaining shuttle flight not directed to ISS, NASA would have had to develop unique procedures and technology because of its unique orbital inclination.

2. A second shuttle would have been necessary to be ready on the launch pad in the event of a problem with the SM4 orbiter. New inspection and the second shuttle technologies would have been required for this one-time mission to HST. Developing these new and unique items and procedures poses a set of risks which, taken individually, are surmountable, but, in the

aggregate, the risks are significantly higher than a shuttle mission to ISS. The total risk, considering the astronauts, the shuttle, the ISS schedule and HST health, I have determined is too high.

O'Keefe further argued that an SM4 mission could not have occurred before June 2006. The unspoken assumption was that in the highly likely event that the RTF was delayed, an SM4 mission would be delayed even further, by which time the HST might be dead. Since a dead HST could not be revived, this raised the possibility that the 18-month training period and effort would be wasted. O'Keefe concluded that "I cannot delegate my ultimate responsibility for decisions related to the safety of human spaceflight to any panel of experts, no matter how distinguished."[43]

Despite his conclusion, after further pressure from Mikulski, two days later O'Keefe asked Admiral Gehman, chairman of the CAIB, to review the matter.[44] Gehman reported back on 5 March, saying that the HST mission was "slightly more risky" than an ISS mission and that it needed "a rich and deep study" to see if it was worth the risk. Mikulski kept up the pressure. In an FY05 budget hearing before the senate subcommittee on 11 March, Mikulski and Subcommittee Chair Kit Bond (Democrat, Missouri) called for the National Academy of Sciences (NAS) and the General Accounting Office to review the risks, costs, and benefits of a shuttle repair mission to the HST. O'Keefe agreed and said he would take such recommendations into account as long as they were not counter to the CAIB report. In a preview of things to come, he further insisted that the NAS charter also include a study of robotic methods for servicing.

A few hours later at a press conference, O'Keefe clearly remained skeptical of a human servicing mission, again citing the risk associated with a rescue mission if something went wrong as fundamentally opposed to the conservative approach to shuttle operations advocated by the CAIB report. But he returned to the idea of a robotic mission, which he now cited as potentially providing new technology in line with the president's vision to send humans to the Moon and Mars.[45]

Congressional pressure came also from the Republicans, notably Senator Kay Bailey Hutchison (Republican, Texas), whose state was the fabled home of the astronauts, the Johnson Space Center. Moreover, she served on the Senate committee that oversaw NASA as well as on the Appropriations Committee that controlled its funding. In a letter to the White House accompanying a petition signed by 26 former astronauts, Hutchison asked President Bush to reconsider early retirement plans for the HST. "The replacement parts needed to keep Hubble operating have already been produced," she wrote. "Should NASA choose robotic transport, installation of parts would be delayed by the expensive and complicated development of mechanical tools. NASA should keep Hubble operational by sending a manned space flight to perform the simple repairs and ensure the satellite's ability to provide crucial knowledge to our space science experts." Similarly, the astronaut petition argued that robotic servicing would have a lower probability of success than a shuttle servicing mission, with only a portion of the tasks accomplished.

The petition had no immediate effect. According to O'Keefe, several of the astronauts later regretted signing it and after discussions with O'Keefe, Senator Hutchison seldom mentioned it. O'Keefe pointed out that the thrust of the petition was that the RTF should occur without complying with the CAIB recommendations. Some astronauts were still of that opinion; O'Keefe was decidedly not. Nevertheless, some of the arguments in the petition later were echoed in the NAS study.[46]

Meanwhile reaction from scientists and the public continued. In early February NASA even had to battle dissent from within, when two reports from anonymous NASA engineers concluded that a mission to the HST was no riskier than the 25 planned missions to ISS.[47] On 11 March Michael Greenfield, the associate deputy administrator for Technical Programs and a member of the Space Flight Leadership Council, wrote a white paper on the HST SM4 for general consumption, attempting to capture all the arguments.[48] Three days later the CBS program *60 Minutes* reported on the HST controversy, an indication of what a cause célèbre the HST had become. In a widely quoted statement in the *60 Minutes* report, Mikulski used a vivid analogy. "The O'Keefe decision is irrevocable, and it's like surgery. If you're going to do an irrevocable decision, you want a second opinion. And that's why I asked for a second opinion—on the risk factors." She would get that second opinion in the form of the NAS study.

Robotic Resolution?

As the reaction to the SM4 cancellation swirled, Frank Cepollina's servicing mission team at GSFC had not been idle. Every day they worked in or near the clean room that contained the HST's new instruments, their fate now uncertain. Serviceable spacecraft had long been a staple at Goddard. Unmanned space missions of the 1960s and 1970s had been considered "failure alley." Of 120 satellites launched during that period, 30% failed to meet mission objectives within 30 days after launch. As a result GSFC started an initiative to develop serviceable spacecraft using modular components. This landmark in spacecraft design proved itself with the Solar Max mission repair in 1984, resulting in extension of the spacecraft lifetime to nine years instead of four. In the years leading up to the first HST Servicing Mission in 1993, spacecraft retrieval had been successful for the Westar and Palapa satellites (1984), Syncom IV (1985), Compton Gamma Ray Observatory (1991), and Intelsat (1992).[49] Cepollina's team had been involved in all four HST servicing missions and was preparing for SM4 when O'Keefe announced its cancellation at the Goddard meeting on 16 January.

For a few days the Goddard team was shocked, but then they went into action. On 20 February, NASA issued a request for information for "HST end of mission alternatives." A month later they were poring over the two dozen ideas submitted. By early April they had produced a plan for a robotic servicing mission. "Our group agreed that a robotic servicing mission of HST conducted in the 2007 timeframe, leveraging existing technologies and robotics expertise stemming from decades of prior investments, and using an approach as general[ly] described herein, falls within the capabilities of NASA and its partners at acceptably low levels of risk," the report stated. It also concluded that its group at GSFC would be best suited to undertake this task.[50]

On 20 April Cepollina and others met with O'Keefe to discuss the robotic mission options. O'Keefe was impressed. The following day at another hearing before Mikulski's Senate subcommittee, he told committee members, "It's looking a lot more promising than I would have told you a few weeks back." While it was not yet clear that a servicing mission would work, O'Keefe said it was his intent to move ahead with a decision on whether or not to implement such a plan by September or October.[51]

The risks of such a plan could not be underestimated. There was confidence that a robot could de-orbit the HST at the end of its life, as planned all along. There was more uncertainty about robotically replacing batteries and gyros, which had never been done before. Most uncertain of all was the ability to robotically install the two new instruments—a difficult task even for experienced astronauts.

At a meeting of the American Astronomical Society in Denver on 1 June 2004, O'Keefe summed up the situation. After reiterating his reasoning for canceling SM4, O'Keefe came to the main point of his address:[52]

> Fortunately, there may be other options for extending the Hubble's useful work...good options that are looking more promising as we've examined them more closely. Our confidence is growing that robots can do the job. For the last few months some of the best and brightest engineers at NASA, within industry and academia have been tirelessly evaluating the options for servicing Hubble by autonomous, robotic means. This

approach now appears to be technically feasible. And the way it started was that we asked the question rather than clinging to a single point solution. Of course on any prospective complex mission of this nature, whether conducted by humans or robots, there are enormous challenges to be faced and no guarantees of success. We are not yet at a point where we have a firm alternative, but we're getting pretty close....In the same can-do spirit that propelled the first Hubble servicing mission, I am very pleased to inform this community that NASA is releasing a call for proposals today for a robotic Hubble servicing capability. This specific request for proposals calls for methods in ascending order of complexity to first, safely de-orbit the Hubble; second, to extend Hubble's service life by adding batteries and new gyroscopes; and third, to install new scientific instruments. This request for proposals is the first step in a long process of developing the best options to save Hubble. We are on a tight schedule to assure a Hubble servicing mission no later than the end of calendar year 2007. But we must act promptly to fully explore this approach. In essence we seek capabilities that highly dexterous robots assisted by humans on the ground could bring to this mission. What we are looking for is not autonomous robotics, but tele-robotics. If this mission goes forward, people will still be servicing Hubble. We are now at a point where these proposals can give us the means to seriously judge whether a robotic servicing mission can be mounted in time to replace Hubble's aging batteries, restore the pointing system with new gyros, and install new scientific instruments.

Thus on 1 June, NASA released a request for proposal for an HST Robotic Servicing–DeOrbit module. The mid-July deadline emphasized the urgency of the task, which had to be accomplished by 2007 before the spacecraft died from lack of power or gyros or both; once dead, it could not be revived. The general idea for servicing was to launch (with a Delta 4 or Atlas 5 rocket) a 20,000 pound (~9,000 kg) two-piece spacecraft that would attach to the HST's aft. It would include robotic eyes, targeting technology, and a robotic arm that would install the new camera, gyros, and spectrograph. The lower part of the spacecraft would then be jettisoned, while the upper section with new batteries and a propulsion system would remain and eventually de-orbit the HST at the end of its life.

At the same meeting O'Keefe also took the opportunity to signal that his reasoning for the original SM4 cancellation had not changed:

> As I have publicly and repeatedly stated, we are committed to implementing the Columbia Accident Investigation Board's recommendations. In their report, the Columbia Accident Investigation Board addressed the need for development of on-orbit inspection, repair, and contingency rescue requirements for every shuttle flight. Those factors bear on any decision to proceed with shuttle operations and acutely bear on requirements surrounding a Hubble servicing mission. A mission to the Hubble would require the development of a unique set of procedures, technologies and tools different from any other mission we'll fly before the shuttle fleet retires. Many of these capabilities which provide safety redundancy for ISS missions are primary or singular for a Hubble mission. Moreover, these Hubble unique methods must be developed and tested promptly before Hubble's batteries and other critical systems give out. We are making steady progress in our efforts to meet the safety requirements for the shuttle return to flight next year. But based on where we are today, prospects are even more challenging than six months ago for our being able to develop in time all required safety and return-to-flight elements for a servicing mission before Hubble ceases to be operational.
>
> The easy route would have been for us to keep plugging along and hope for the best. But "hope" is not a management method we should rely on to keep Hubble operating. The Columbia Accident Investigation Board recommended that we change our culture to a commitment to "prove that it is safe" rather than place the burden of proof on folks to "prove that it's not safe." Well, with that guidance in mind, we're nowhere near close to proving that it's safe. It's not the unknowns we are wrestling with,

it's the knowns that we haven't yet devised a way
yet to conquer. Accordingly, I found it would not be
responsible to prepare for a servicing mission, only to
find that the required actions identified by the board
could not be implemented. This likely condition would
pose a Hobson's choice. It is likely we would have
two untenable alternatives to choose from. Either
fly the Hubble mission without fully complying with
the board's recommendations or allow Hubble to
simply cease to function. The prospect of either of
these options if we had put all our eggs in the shuttle
servicing basket is simply unacceptable. Equally
untenable is the expectant atmosphere that would exist
all the way up to a launch "go or no go" decision. This is
precisely the type of "schedule pressure" that the board
quite correctly stated would significantly undermine
the future safe operation of the shuttle.[53]

This reasoning catapulted robotic servicing to the forefront. Unfortunately, the NAS report would not let O'Keefe off the hook quite so easily.

The National Academy Report

On 16 April, in response to the Mikulski–Bond request, the NAS announced the members of its Committee on Assessment of Options for Extending the Life of the HST. Chaired by Bell Laboratories consultant Louis Lanzerotti, the NAS panel met with Administrator O'Keefe on 22 June and issued an interim report on 13 July. The interim report, issued because of the urgency of the HST situation in a letter to O'Keefe, included three findings and recommendations:[54]

FINDING. Compelling scientific returns will
result from a servicing mission to the Hubble Space
Telescope that accomplishes the scientific objectives of
the originally planned NASA servicing mission SM-4.
RECOMMENDATION. The committee urges that
NASA commit to a servicing mission to the Hubble
Space Telescope that accomplishes the objectives of
the originally planned SM-4 mission, including both
the replacement of the present instruments with
the two instruments already developed for flight—
the Wide Field Camera-3 and the Cosmic Origins

Spectrograph—and the engineering objectives, such as gyroscope and battery replacements. Such a servicing mission would extend the life of this unique telescope and maximize its productivity. Other potential options to extend the useful life of Hubble—for example, by servicing components such as batteries and gyroscopes but without replacing instruments—will be studied by the committee as part of its charge. However, such a reduced level of servicing has not been featured in the repair strategies that the committee has heard about to date.

FINDING. The proposed Hubble robotic servicing mission involves a level of complexity, sophistication, and technology maturity that requires significant development, integration, and demonstration to reach flight readiness. RECOMMENDATION. As an early step, NASA should begin immediately to take an active partnership role that includes HST-related demonstrations in the robotics space experiments that are now under way in other agencies in order to ensure that the returns from these experiments can be beneficial to a potential robotic Hubble servicing mission.

FINDING. Because of inherent uncertainties in the early stages of development of a robotic mission to the Hubble Space Telescope, as well as the uncertain current status of the shuttle return-to-flight program, the key technical decision points for committing to a specific service scenario are at least a year in the future. RECOMMENDATION. At the same time that NASA is vigorously pursuing development of robotic servicing capabilities, and until the agency has completed a more comprehensive examination of the engineering and technology issues, including risk assessments related to both robotic and human servicing options, NASA should take no actions that would preclude a space shuttle servicing mission to the Hubble Space Telescope.

Democratic Senator Mikulski and Republican Representative Sherwood Boehlert, chair of the House Science Committee, endorsed the recommendations, demonstrating bipartisan support for extending the HST's lifetime. O'Keefe's reaction was also positive: "We agree with the committee's view that the Hubble Space Telescope is arguably the most important telescope in history. NASA is committed to exploring ways to safely extend the useful scientific life of Hubble. The challenges of a robotic mission are under examination and we'll continue our exhaustive and aggressive efforts to assess innovative servicing options. In parallel with the committee's ongoing research and deliberations, NASA will evaluate proposals we expect to receive shortly. Along the way, we'll keep options open to assure the best possible outcome." But the bottom line of the report was clear, and a potential headache for Administrator O'Keefe: while it should pursue the robotic option, the panel recommended that NASA should not preclude using the shuttle for a servicing mission along the lines originally proposed before the *Columbia* accident, enabling the full servicing with new instrumentation. This was the option that O'Keefe thought he had ruled out six months earlier.

By December 2004 another problem with the robotic option had come to the fore: cost. The Aerospace Corporation, commissioned by NASA to study all options, concluded in a report dated 3 August that a robotic servicing mission was "high risk," would cost 2 billion dollars, and would take five years to implement, by which time the HST would be dead. The most certain way to a successful servicing mission, they concluded, was using the method used four times before: the shuttle. Another option would be to place the new instruments on a new bare-bones telescope. An executive summary of the report was not available to the public until 6 December.[55]

Two days later, on 8 December, the NAS released its final detailed 147-page congressionally mandated report. The findings and recommendations were a rebuff to the robotic mission in every way. The committee found that the technology for robotic servicing required a level of complexity that was inconsistent with the 39-month development schedule needed to reach the telescope in time, even with the expertise of the Goddard team, which had little experience in autonomous rendezvous and docking. It found further that meeting the CAIB and NASA requirements relative to inspection and repair, safe haven, and shuttle rescue was viable; that "the shuttle crew safety risks of a single mission to ISS and a single HST mission are similar and the relative risks are extremely small"; and that space shuttle crews and their ground-control teams had a proven track record of developing innovative techniques in orbit—likely to be needed for an HST servicing mission. And, they found that the HST was a uniquely powerful instrument worthy to be saved.

The three final recommendations could not have been clearer—or more diametrically opposed to O'Keefe's original intention to cancel a shuttle servicing mission:

1. The committee reiterates the recommendation from its interim report that NASA should commit to a servicing mission to the Hubble Space Telescope that accomplishes the objectives of the originally planned SM-4 mission.

2. The committee recommends that NASA pursue a shuttle servicing mission to HST that would accomplish the above stated goal. Strong consideration should be given to flying this mission as early as possible after return to flight.

3. A robotic mission approach should be pursued solely to de-orbit Hubble after the period of extended science operations enabled by a shuttle astronaut servicing mission, thus allowing time for the appropriate development of the necessary robotic technology.[56]

Six months after O'Keefe had committed to studying the robotic options for an HST rescue, the option seemed to be closed by a blue ribbon panel. Yet the decision remained for O'Keefe to make. Less than a week later, however, on 13 December, O'Keefe resigned as NASA administrator. There was no clear cause-and-effect relationship as far as the HST was concerned. His resignation letter to President Bush cited family reasons, and the desire to accept the chancellorship at Louisiana State University, at more than three times his civil service salary ($500,000 versus $158,000). It was also a few weeks after Congress had given NASA a record $16.2 billion appropriation, which O'Keefe had shepherded through Congress and took as a mandate for the new vision for space exploration enunciated by President Bush the previous January. Two days later, on 16 January, O'Keefe had announced the cancellation of the HST servicing mission, setting in motion the remarkable chain of events described in this essay.

Almost a year after his HST decision was announced, and a few weeks before his resignation announcement, O'Keefe still believed he had made the right decision for the HST. "I'm not sure 'vindicate' is the right word, because it's a long time before that would be demonstrated. But I certainly feel that as passing time has gone by that it was the right call. Painful as it was, difficult as it was, I don't have any reservations about it at all. I think it has helped to prompt a whole different way of looking at the problem." He viewed his decision as essential to shifting NASA's culture, and believed that the critics of his decision had submerged "adherence to principle, to a set of objectives necessary to demonstrate the credibility of this agency to do what we say we're going to do, that we're committed to do and actually prove that we are going to do it. That credibility, by the way, has always been much in question, always been much accused of being not quite as rigid as it should be. It's a point I find to be really problematic, and we have a real challenging kind of history in this case."[57]

Summary and Lessons Learned

As of the end of 2004 NASA was involved in a race not of its choosing—to adopt the best option to service the HST before it ceased operations in 2007 or 2008 or to decide to de-orbit the spacecraft at some point after its death. Whether a servicing mission would be carried out with robots or humans remained an open question. Robots had no record of capability to accomplish the simple tasks of replacing batteries and gyros, much less the far more complicated tasks of installing new instruments. Humans had a proven record of servicing with the Space Shuttle, but the Space Shuttle might not be able to make it in time. At the core of the matter was an assessment of the relative risk of a shuttle HST mission compared with a shuttle ISS mission. This assessment remained controversial, with the NAS panel concluding that the risk differential was "extremely small," and O'Keefe maintaining that his responsibility to crew safety and the CAIB recommendations precluded a shuttle flight in the time remaining. Both options required long lead times, meaning a decision needed to be made soon. Ironically, a telescope that had been the brunt of jokes when it was first launched with its mirror problems now had triggered a national outpouring of concern, as Congress, scientists, and the general public clamored for its life to be extended. It was truly a remarkable turn of events in the history of space science.

What made the controversy even more remarkable was the intimate connection with human spaceflight. The fortunes of the HST had always been tied to the Space Shuttle. Its launch was delayed by the shuttle *Challenger* accident, and its servicing made possible by four remarkable shuttle missions. Critics who had at times portrayed the Space Shuttle as an expensive toy stuck in Earth orbit had been partially mollified by the HST servicing missions, which had demonstrated a place for both humans and robots in space and kept a steady stream of inspiring astronomical images beaming toward Earth. Now, its last servicing mission was threatened because of the shuttle *Columbia* accident. To this extent space science and human spaceflight had become intimately wedded, and the reputations of both were at stake.

In short, the HST servicing mission controversy was part of an American space program at a crossroads at the beginning of a new millennium. It should be remembered that it was only a small part of NASA's portfolio, given NASA's far flung concurrent space science programs such as the Mars Exploration Rovers, the Cassini–Huygens spacecraft, and the elements of the new exploration vision that were being discussed as the HST controversy raged. Nevertheless, the public, scientific, and congressional reaction to the *Columbia* accident, and the subsequent controversy that has swirled around the SM4 cancellation decision and the related RTF, raised issues that went to the core of NASA's mission, and indeed held relevance to high-tech agencies beyond NASA. Among them are the following.

ORGANIZATIONAL LEARNING: NASA had long been criticized for not being a learning organization, specifically for not learning the lessons of *Challenger*. Yet, when the agency took the lessons to heart in the wake of the *Columbia* accident—with serious implications of which the HST was only one—it was subjected to withering criticism. Associate Administrator Ed Weiler summarized the irony succinctly, "Sean feels he is following Admiral Gehman's report. Sean has been beaten up by Congress. He has been beaten up by the press for not having a safety culture at NASA. He has made a safety decision. I'll leave it to the reader."[58] Many NASA managers felt that the recommendations of the CAIB report were all too soon forgotten by a Congress protecting its constituency, scientists more interested in science than human risk, and a public that wanted nice pictures. How does one weigh the agendas of those constituencies against the risk involved? In answering that question it should be obvious that managers need to be able to make decisions that are not always the popular choice based on a hoped-for outcome. In O'Keefe's words, "Hope cannot be used as a management tool."

REACTION TO EXECUTIVE DECISIONS: Asked for lessons learned from the HST case, O'Keefe stated, "first and foremost don't be intimidated by the negative reaction. And if you make a decision exclusively on the popularity of it you may well likely make a poor decision. And so the lesson learned from this one is stay the course, hold on, the wind is going to get pretty wild. And the hurricane force blow may be a little bit violent at times, but it will eventually get you through to where you need to go. And if you fold up you may end up doing the popular choice, but you may not be doing the right thing."[59]

COMMUNICATION OF DECISIONS: Decisions are inevitably more complex than they appear to the media, the public, and even those relatively close to those decisions inside an agency. In this case the HST SM4 cancellation appeared to outsiders to be a budget issue. The unanimous claim inside NASA, from the administrator to the comptroller to the HST managers was that it was not a budget issue; a claim borne out by the evidence but that never seemed to penetrate the media. Similarly, the servicing mission cancellation, having been announced two days following the new space exploration vision, was immediately identified as a casualty of that vision. In fact, we have seen that the decision to cancel SM4 was made in late November 2003 and reflected in budget documents by 2 December, six weeks prior to the president's announcement. Although the vision announced on 14 January 2004 had been in the making for several months, there is no evidence that HST had been part of those deliberations. Communicating facts both internal and external to the agency is essential but sometimes futile given human failings. In order to set the record straight, it is essential that objective histories of policy decisions be encouraged, written, and widely disseminated and that the lessons of history be learned. To the extent possible, objective history needs to be taken into account before decisions are made.

PLANNING FOR THE UNPLANNED: To paraphrase Robert Burns, "The best-laid plans of mice and men often go awry." That was certainly the case with the timing of the SM4 cancellation announcement, which had been carefully planned for late January after all constituencies could be properly briefed but which was in actuality dictated by a leak to the press. This is such a common event in Washington that it should itself be taken into account in planning.

SCHEDULE PRESSURE: Schedule pressure is the norm at NASA, and at many other agencies, but unlike most other agencies at NASA it could be seen as a contributing factor to spectacular and very public national tragedy. Nevertheless, schedule pressure is not necessarily bad. As O'Keefe said, "there is good schedule pressure and there is bad schedule pressure. Bad schedule pressure is when you've got everybody cutting corners in order to do something and risking people's lives to do it. Good schedule pressure is something where there's an imperative out there to come to solution, come to closure, reach understanding, find a consensus, so that you can get on with the task constructively so that it's just not lingering forever and ever. But in the process the risk should be not at the expense of the potential of someone surviving the experience or not." Navigating a project or an agency between good schedule pressure and bad schedule pressure is perhaps one of the most difficult, but necessary, tasks that managers face.

RISK AVERSION: The HST SM4 controversy and the schedule for RTF raise important questions about risk aversion, and whether the USA has become a risk-averse society. If anything good has come out of the *Columbia* accident and the HST SM4 cancellation controversy, it is perhaps that a more sophisticated discussion of risk aversion has been generated. O'Keefe noted, "The bulk of the folks who really are dedicated to this, who care about it, really are coming to grips with the fact that this really is a Hobson's choice. This really is coming to grips with that intellectual challenge of how you look at problems and issues, and how you ultimately then have to make decisions about waiving what you believe to be principles, or not, or sticking with them and having it be a very withering circumstance based on what is a very difficult definition of what is risk."[60]

The question bears on situations far beyond NASA, and on the role of the USA as a creative society that does not shrink from exploration. "What's acceptable risk?" O'Keefe asked. "Some people sat back and said, you know, this is indicative of the fact that the American public can't stand losses. I think that's wrong. I think that's a different case, I think we're more tolerant and accepting of risks quite frankly than many other cultures and many other civilizations throughout the course of history in some ways." One might ask why not just let the astronauts go, since there would certainly be no lack of volunteers. Management's answer is that one does not make policy based on daredevils: "It's not a question of whether or not the people who are exposed to it are witting of that level of risk, it should be a question of whether or not we as the individuals who are responsible for the conduct of those operations are witting of that risk on their behalf, and have done our diligent best to avert as much of the challenge and the unknowns about it as we possibly can. We'll never eliminate the risk."[61]

O'Keefe considers the risk question one of the most important results to come out of the current controversy.

> And so it's kind of gotten to that level of sophistication
> of the debate which is interesting and really, really
> good. It is a more positive way; I think a more
> constructive way, to evaluate what it is we are willing
> to accept here. And it isn't the yee-haw school of get
> aboard and fly anytime you want nor is it the other
> side, which is asserted that some would argue that I'm
> advocating here which is aversion, it isn't that either.
> It's saying there is a metric, there's a measurement,
> there's a standard by which you can measure what
> you're prepared to accept, and that was set for us in a
> 248 page report, with seven volume set of appendices
> that went with it. Produced lots of details, that's a good
> way to measure that to say this is what the standard of
> risk acceptance ought to be for this program. But in
> a broader sense, it also is calling into question what is
> the acceptance of risk that we as a people, as a culture,

will adhere to in pursuit of exploration objectives, science pursuit, whatever. And I think frankly the answer to that is, it is pretty tolerant. People are pretty tolerant. If we weren't, the entire space flight program would have shut down on 2 February 2003. And it would have had another opportunity to shut down on 27 August, the day after the report was released, and we didn't. Each of these steps turns on, again, I think, a deeper reflection of these things which again is coming out in the public debate more and more evidently—a willingness of acceptance of risk as long as you understand what the objectives are all about and whether those are objectives we think are worth encountering that level of risk. And that there is diligence exercised along the way to mitigate it as much as possible, that's an expectable standard. Boy, that's a much more sophisticated debate and argument that we've had about this in a long time.[62]

Reversal of Fortune

In the wake of Sean O'Keefe's resignation effective 11 February 2005, the hopes of those who advocated a final HST servicing mission rested on Michael D. Griffin, who arrived as the new NASA administrator on 13 April. Griffin, an engineer who had worked as NASA's associate administrator for Exploration in the early 1990s, came to the administrator's position from the Johns Hopkins University Applied Physics Laboratory. His résumé boasted years of experience in both space science and human spaceflight, but he left no doubt that his top priority was to implement the human spaceflight program, known as Constellation, that President George W. Bush had enunciated in January 2004.[63]

How Griffin would act on the HST was an open question. But he indicated his thinking already during his Senate confirmation hearings on 12 April 12, 2005—24 years to the day after the first shuttle flight. "I would like to take the robotic mission off the plate," he stated, bowing to the NAS report that had concluded it was unfeasible. "I believe the choice comes down to reinstating a shuttle servicing mission or possibly a very simple robotic de-orbiting mission." The latter referred to de-orbiting the HST for safety reasons sometime after 2020, without having performed the risky servicing mission—in other words, long after it was dead. That decision was hardly urgent, but the shuttle servicing mission was. "When we return to flight it will be with essentially a new vehicle, which will have a new risk analysis associated with it. At that time I think we should reassess the earlier decision in light of what we learn after we return to flight."[64] The essential ingredient was a successful RTF. That was accomplished when shuttle *Discovery* was launched on 26 July 2005. Two more successful flights followed in July and September the following year.

In the wake of three successful shuttle missions after the RTF, on 31 October 2006—33 months after O'Keefe had announced the cancellation of SM4—Administrator Griffin announced to employees at the same venue (GSFC) that the fifth servicing mission (still known as SM4) would indeed take place. In the official press release from NASA Headquarters Griffin was quoted as saying, "We have conducted a detailed analysis of the performance and procedures necessary to carry out a successful Hubble repair mission over the course of the last three shuttle missions. What we have learned has convinced us that we are able to conduct a safe and effective servicing mission to Hubble. While there is an inherent risk in all spaceflight activities, the desire to preserve a truly international asset like the Hubble Space Telescope makes doing this mission the right course of action."[65] At a press conference the same day, Griffin noted, "I don't believe I have talked to anyone in the agency, from flight crew to flight ops manager to even budget guys...who thinks we shouldn't do this." There was not, however, unanimity from the astronaut office or the engineers about "launch on need," the decision to have a second shuttle ready to go on launch pad 39B for rescue in case of a problem at the HST. In any case, the second "rescue" shuttle would be in place at the time of the servicing mission launch.[66]

The decision to return to the HST was hailed by astronomers, Congress, and the public alike. Senator Barbara Mikulski, the long-time HST supporter whose district included the Goddard center, enthused, "This is a great day for Maryland, for America, but most of all, for science. Hubble is a national asset and a national priority. Without question, Hubble has been the most successful NASA program since Apollo. And like Apollo, Hubble has helped America lead the way in discovery and exploration."[67]

The SM4 was originally scheduled for 11 September 2008 on shuttle *Atlantis* (STS-125). But due to a malfunction of the HST's control unit–science data formatter, which affected the storage and transmittal of all science data to Earth, the mission was postponed until 2009 while a replacement unit was checked out. On 11 May 2009 SM4 was finally launched aboard the *Atlantis* five years after originally scheduled. Among the astronauts were mission specialists John Grunsfeld and Michael Massimino, both veteran spacewalkers who had serviced the HST before, and rookie astronauts Andrew Feustel and Michael Good. Over a series of five spacewalks from 14–18 May the four astronauts accomplished all their goals, though not without challenges. They installed two new instruments: the Wide Field Camera 3 (WFC3) and the Cosmic Origins Spectrograph (COS), the latter used exclusively in the ultraviolet, increasing the HST's ultraviolet sensitivity at least ten times. In addition to these two new instruments, the astronauts also repaired two instruments, the Advanced Camera for Surveys (ACS) and the Space Telescope Imaging Spectrograph (STIS). Since a power failure in 2004, STIS had been dysfunctional; the ACS had suffered an electrical short in 2007. The astronauts also replaced the 18-year old batteries and installed six new gyroscopes and a new Fine Guidance Sensor. Finally, they installed the spare Science Instrument Command and Data Handling Unit, the apparatus that had caused the eight-month launch delay. All told, SM4 was the heaviest servicing mission of all, carrying 22,000 pounds (~10,000 kg) of hardware to the HST. After the mission, the HST was at the apex of its performance.

With its new lease on life the HST was projected to finish its remarkable mission around the 2012 to 2015 timeframe. As John Grunsfeld remarks in his essay in this volume, "Our warranty is three years, labor not included. Five years is totally reasonable. Beyond that is something we'd be delighted to have, especially overlapped with the James Webb Telescope." While the latter is looking unlikely due to delays, 25 years of cutting-edge astronomy would be a remarkable legacy for a telescope once given up for dead.

How was the HST's fifth servicing mission cancelled by one administrator only to be green-lighted by his successor? As we have seen, O'Keefe's reluctance can only be understood in the context of the *Columbia* accident and the report of the Columbia Accident Investigation Board. Griffin's decision can only be seen in the context of his belief that exploration, human or robotic, is among the most important things humans can do.[68] Risk assessments were involved in both decisions, and despite the technical evaluations, there remained a large measure of subjectivity in both cases. Griffin's gutsy decision was more in tune with the idea that safety is the second priority in any bold adventure; having taken all precautions, the first priority is to go, otherwise no explorers would ever have left home.[69]

Acknowledgments

I thank all those interviewed, as well as the NASA History Office staff, Harry Lambright for discussions, and Retha Whewell for verifying dates of crucial meetings and other assistance.

Notes

1   In the immediate aftermath of his decision to cancel SM4 for the HST, NASA Administrator Sean O'Keefe requested this independent study be undertaken by the author, in his role as NASA chief historian, in order to document in detail the events that led to the cancellation decision. This report was completed 17 December 2004 and is published here for the first time. It is essentially unchanged except for the final section ("Reversal of Fortune") added in 2012.

2   Robert W. Smith, *The Space Telescope: A Study of NASA, Science, Technology, and Politics* (Cambridge: Cambridge University Press, 1989).

3   John M. Grunsfeld, Oral History Interview (OHI), 13 April 2004, p. 2, NASA Historical Reference Collection, NASA Headquarters, Washington, DC. All OHIs cited in this essay are archived at this location, including both tapes and transcripts.

4   Grunsfeld, OHI, 13 April 2004, p. 5.

5   Columbia Accident Investigation Board, *"Columbia Accident Investigation Board Report*, volume 1, Appendix A, "The Investigation," p. 231. The full report is online at http://caib.nasa.gov/.

6   Columbia Accident Investigation Board, *Columbia Accident Investigation Board Report*, 225.

7   Return to Flight Task Group, *NASA's Implementation Plan for Return to Flight and Beyond*, 8 September 2003. This final report may be found at http://govinfo.library.unt.edu/returnflight/reports/final_report.html.

8   David Black (Chair), Alan Dressler, Robert Kennicutt, George Miley, William Oegerle, Glenn Schneider and J. Michael Shull, "Final Report of the Hubble Space Telescope Post-SM4 Scientific Review Panel," 25 April 2003, available online at http://nicmosis.as.arizona.edu:8000/REPORTS/Post-SM4_Final_Report.pdf.

9   John Bahcall, Barry Barish, Jacqueline Hewitt, Christopher McKee, Martin Rees, and Charles Townes, "Report of the HST-JWST Transition Panel," 14 August 2003.

10 Sean O'Keefe, OHI, 22 April 2004, pp. 1–2.

11 William F. Readdy, OHI, 23 March 2004, p. 6.

12 Ed Weiler, OHI, 2 Feb 2004, pp. 12, 13; briefing slides, "HST SM4 Options Briefing with Mr. O'Keefe, 9:30 a.m., 7 November 2003," NASA History Office. In addition to O'Keefe, present at the 7 November meeting were Isakowitz, Readdy, O'Connor, and Chief of Staff John Schumacher.

13 Steve Isakowitz, OHI, 18 Feb 2004, pp. 10, 19, 25, 28.

14 Isakowitz, OHI, 18 Feb 2004, pp. 10, 19, 25, 28.

15 Isakowitz, OHI, 18 Feb 2004, pp. 10, 19, 25, 28; O'Keefe, OHI, 29 October, 2004, pp. 2, 4.

16 O'Keefe, OHI, 22 April 2004, pp. 3–5; Weiler, OHI, 2 February, 2004, pp. 4–7, 15; Isakowitz, OHI, 18 Feb 2004, pp. 19, 27–28.

17 Weiler, OHI, 2 Feb 2004, pp. 15, 16.

18 Remarks by Administrator O'Keefe at the American Astronomical Society Annual Meeting, Denver, Colorado, 1 June 2004.

19 The minutes of the 9–11 December Johnson Space Center meeting are found at http://govinfo.library.unt.edu/returnflight/reports/final_report.html.

20 O'Keefe, OHI, 22 April 2004, pp. 3–6. The shuttle return to flight date of March 2005 was officially announced 19 February 2004. It was later moved to May 2005.

21 Grunsfeld, OHI, 13 April 2004, p. 7.

22 Grunsfeld, OHI, 13 April 2004, p. 11.

23 Brian O'Connor, OHI, 1 November 2004, p. 9.

24 O'Connor, OHI, 1 November 2004, pp. 15, 16.

25 Grunsfeld, OHI, 13 April 2004, pp. 13, 14; O'Keefe, OHI, 29 October, 2004, pp. 13, 14.

26 Details of NASA–White House interaction in planning the new space vision are given in Frank Sietzen Jr. and Keith L. Cowing, *New Moon Rising: The Making of American's New Space Vision and the Remaking of NASA* (Burlington, Canada: Collector's Guide Publishing, 2004).

27 Isakowitz, OHI, 18 Feb 2004, p. 21.

28 NASA, *The Vision for Space Exploration* (February 2004), online at http://www.nasa.gov/pdf/55583main_vision_space_exploration2.pdf; *Science* cover story and series of articles "Remaking NASA," *Science* 303 (30 Jan 2004):610–619, and editorial 589; Jeffrey Kluger, "Mission to Mars," *Time* 23 January 2004:42–52.

29 Kathy Sawyer, "Visions of Liftoff, Grounded in Political Reality," *Washington Post* 15 January 2004, A1; Isakowitz, OHI, 18 Feb 2004, pp. 26, 27; Grunsfeld, OHI, 13 April 2004, p. 19.

30 Grunsfeld, OHI, 13 April 2004, p. 20.

31 Michael Moore, OHI, 23 January 2004, p. 36; Jennifer Wiseman, OHI, 29 January 2004, p. 15.

32 Wiseman, OHI, 29 January 2004, p. 19. Steven Beckwith, "Space Telescope Science Institute Status Report," 22 January 2004. This report by Beckwith, STScI Director, describes the 16 January meeting in considerable detail.

33 Isakowitz, OHI, 18 Feb 2004, p. 16; Readdy, OHI, 23 March 2004, p. 14.

34 Kelly McBride, "Hubble, the Beloved: Decision to Stop Maintenance of Telescope Generates Outpouring," *Washington Post* 29 January 2004, A27.

35 Jeffrey Kluger, "Mission to Mars," *Time* 26 January 2004, 42–52.

36 O'Keefe, OHI, 22 April 2004, pp. 9–10.

37 O'Keefe, OHI, 22 April 2004, p. 11.

38 American Astronomical Society Exploder to members, 28 January 2004. [Editor's note: Beckwith's assertion does not bear out historical scrutiny.]

39 Weiler, OHI, 2 Feb 2004, p. 17.

40 *Physics Today*, March 2004, 29.

41 Andrew Lawler, "Scientists Add up Gains, Losses, in Bush's New Vision for NASA," *Science* 303 (23 January 2004), 444–445.

42 Mikulski to O'Keefe, 21 January 2004; O'Keefe to Mikulski, 26 January 2004.

43 O'Keefe to Mikulski, 26 January 2004.

44 Brian Berger, "O'Keefe Asks CAIB Chairman to Review Hubble Decision," *Space News* 2 February 2, 2004, 7.

45 Andrew Lawler, "Academy, GAO to Study Possible Robotic Hubble Mission," *Science* 303 (19 March 2004), 1745.

46 Letter from Senator Hutchinson to President Bush dated 26 May 2004, with accompanying astronaut petition.

47 "Applicability of CAIB Findings/Recommendations to HST Servicing," and "Risk Considerations for HST and ISS Flights."

48 Michael Greenfield, "Cancellation of the Fifth (SM-4) Hubble Servicing Mission," 11 March 2004, NASA History Office.

49 Frank Cepollina, OHI 2 May 2004, 6ff; Cepollina, "On Orbiting Servicing" presentation 17 July 2004.

50 Goddard Space Flight Center, "Hubble Space Telescope Robotic Servicing Mission: Mission Feasibility Study for the WFC3 and COS Scientific Instrument Installation," 6 April 2004.

51 "O'Keefe: Robotic Options for Saving Hubble Promising," *Space News* 26 April 2004, 7.

52 Remarks by Administrator O'Keefe at the American Astronomical Society Annual Meeting, Denver, Colorado, 1 June 2004.

53 O'Keefe, remarks to American Astronomical Society, 1 June 2004. O'Keefe elaborates on the Hobson's choice in OHI, 29 October, 2004, pp. 9, 10.

54 National Academy of Sciences committee letter to O'Keefe, "Assessment of Options for Extending the Life of the Hubble Space Telescope: Letter Report," 13 July 2004.

55 Aerospace Corporation, "Hubble Space Telescope (HST) Servicing Analyses of Alternatives (AoA): Final Delivery," Executive Summary, 3 August 2004. John Kelly, "Study: Robotic Repair Mission to Hubble Too Costly," *Florida Today* 7 December 2004.

56 National Research Council, Committee on the Assessment of Options for Extending the Life of the Hubble Space Telescope, *Assessment of Options for Extending the Life of the Hubble Space Telescope: Final Report*, National Academy of Sciences, 2005.

57 O'Keefe, OHI, 29 October, 2004, pp. 5, 10, 11.

58 Weiler, OHI, 2 February 2004, p. 14.

59 O'Keefe, OHI, 29 October, 2004, p. 26.

60 O'Keefe, OHI, 22 April 2004, pp. 20–23.

61 O'Keefe, OHI, 22 April 2004, pp. 20–23

62 O'Keefe, OHI, 22 April 2004, pp. 20–23.

63 W. Henry Lambright, "Launching a New Mission: Michael Griffin and NASA's Return to the Moon," online at http://www.spacepolicyonline.com/images/stories/Lambright-MoonReport1.pdf, (accessed 1 February 2012).

64 Robert Zimmerman, *The Universe in a Mirror: The Saga of the Hubble Space Telescope and the Visionaries Who Built It* (Princeton, N.J.: Princeton University Press, 2008), 206–207.

65 "NASA Approves Mission and Names Crew for Return to Hubble," NASA Headquarters Press release, 31 October 2006, online at http://www.nasa.gov/home/hqnews/2006/oct/HQ_06343_HST_announcement.html (accessed 1 February 2012).

66 "NASA Press Conference—Shuttle Return to Hubble—Transcript," 31 October 2006, 11, and online at http://www.spaceref.com/news/viewsr.html?pid=22474 (accessed 1 February 2012).

67 "Mikulski Applauds Hubble Announcement, Says Decision is Right for America," SpaceRef, 1 November 2006, online at http://mikulski.senate.gov/media/record.cfm?id=265518 (accessed 26 January 2012).

68 Griffin's emphasis on exploration is clear throughout: Michael Griffin, *Leadership in Space: Selected Speeches of NASA Administrator Michael Griffin, May 2005–October 2008* (Washington: NASA Headquarters, 2008). It is also clear in his address at NASA's 50th anniversary symposium in October 2008: Michael Griffin, "NASA at 50," in *NASA's First 50 Years: Historical Perspectives*, ed. Steven J. Dick (Washington: NASA SP-2010-4704, 2010), 1–9, online at http://history.nasa.gov/SP-4704.pdf.

69 Steven J. Dick and Keith Cowing, eds., *Risk and Exploration: Earth, Sea and the Stars* (Washington: NASA SP-2005-4701, 2005), online at http://history.nasa.gov/SP-4701/riskandexploration.pdf. Sean O'Keefe convened this "NASA Administrator's Symposium" in the wake of the *Columbia* accident to discuss these issues in a broad historical context.

Select Bibliography

History of Cosmology

Berendzen, Richard, Richard Hart, and Daniel Seeley. *Man Discovers the Galaxies*. New York: Science History Publications, 1976.

Boss, Alan. *Looking for Earths: The Race to Find New Solar Systems.* New York: John Wiley and Sons, Inc., 1998.

Christianson, Gale E. *Edwin Hubble: Mariner of the Nebulae*. New York: Farrar, Straus, Giroux, 1995.

Clair, Jean, ed. *Cosmos: From Romanticism to the Avant-garde*. New York: Prestel, 1999.

Crowe, Michael J. *Modern Theories of the Universe: From Herschel to Hubble*. New York: Dover, 1994.

DeVorkin, David H., ed. *Beyond Earth: Mapping the Universe*. Washington, D.C.: National Geographic, 2002.

Dick, Steven J. *Discovery and Classification in Astronomy: Controversy and Consensus*. New York: Cambridge University Press, 2013.

Goodwin, Simon. *Hubble's Universe: A Portrait of our Cosmos.* New York: Viking Penguin, 1997.

Gott, Richard J., and Robert J. Vanderbei. *Sizing Up the Universe: A New View of the Cosmos*. Washington, D.C.: National Geographic, 2010.

Gribbin, John. *The Birth of Time: How Astronomers Measured the Age of the Universe.* New Haven, Conn.: Yale University Press, 1999.

Hetherington, Norriss S., ed. *Encyclopedia of Cosmology: Historical, Philosophical, and Scientific Foundations of Modern Cosmology*. New York: Garland Publishing, 1993.

Kirshner, Robert P. *The Extravagant Universe: Exploding Stars, Dark Energy, and the Accelerating Cosmos*. Princeton, N.J.: Princeton University Press, 2002.

Kragh, Helge. *Conceptions of Cosmos from Myths to the Accelerating Universe: A History of Cosmology*. New York: Oxford University Press, 2006.

Learner, Richard. *Astronomy through the Telescope*. New York: Van Nostrand Reinhold, 1981.

Leverington, David. *New Cosmic Horizons: Space Astronomy from the V-2 to the Hubble Space Telescope*. New York: Cambridge University Press, 2001.

Levin, Frank. *Calibrating the Cosmos: How Cosmology Explains Our Big Bang Universe*. Chichester, UK: Springer-Praxis, 2007.

North, John. *Cosmos: An Illustrated History of Astronomy and Cosmology*, revised edition. Chicago: University of Chicago Press, 2008.

———. *The Norton History of Astronomy and Cosmology*. New York: W.W. Norton and Co., 1994.

Nussbaumer, Harry, and Lydia Bieri. *Discovering the Expanding Universe*. New York: Cambridge University Press, 2009.

Overbye, Dennis. *Lonely Hearts of the Cosmos: The Story of the Scientific Quest for the Secret of the Universe*. Boston: Back Bay Books, 1999.

Panek, Richard. *Seeing and Believing: How the Telescope Opened Our Eyes and Minds to the Heavens*. New York: Viking, 1998.

Petersen, Carolyn Collins, and John C. Brandt. *Visions of the Cosmos*. New York: Cambridge University Press, 2003.

Rees, Martin. *New Perspectives in Astrophysical Cosmology*. New York: Cambridge University Press, 2002.

Sandage, Allan. *Centennial History of the Carnegie Institution of Washington, Volume 1. The Mount Wilson Observatory: Breaking the Code of Cosmic Evolution*. Cambridge, UK, and New York: Cambridge University Press, 2004–2005; reprinted January 3, 2013.

Sandage, Allan, and John Bedke. *Atlas of Galaxies: Useful for Measuring the Cosmological Distance Scale*. Washington, D.C.: Scientific and Technical Information Division, National Aeronautics and Space Administration. For sale by the Superintendent of Documents, U.S. Government Printing Office, 1988.

Sharov, Alexander S., and Igor D. Novikov. *Edwin Hubble, the Discoverer of the Big Bang Universe*. Cambridge, UK: Cambridge University Press, 2005.

Singh, Simon. *Big Bang: The Origin of the Universe*. New York: Fourth Estate, 2005.

Smith, Robert W. *The Expanding Universe: Astronomy's "Great Debate," 1900–1931*. New York: Cambridge University Press, 1982.

Tucker, Wallace, and Karen Tucker. *The Cosmic Inquirers: Modern Telescopes and Their Makers*. Cambridge, Mass.: Harvard University Press, 1986.

Wilford, John Noble, ed. *Cosmic Dispatches: The New York Times Reports on Astronomy and Cosmology*. New York: W.W. Norton and Co., 2001.

Wright, Helen. *The Great Palomar Telescope*. London: Faber and Faber, 1953.</refs>

Hubble History

American Institute of Aeronautics and Astronautics (AIAA). *Large Space Telescope—A New Tool for Science: [Proceedings] AIAA 12th Aerospace Sciences Meeting, Washington, D.C., January 30–February 1, 1974.* Edited by F. Pete Simmons. Reston, Va.: AIAA, 1974.

Bahcall, J. N., and L. Spitzer, Jr. "The Space Telescope." *Scientific American*, 247 (1982):38–49.

Boggess, Albert, and David S. Leckrone. "The History and Promise of the Hubble Space Telescope." *Optics & Photonics News,* 1(3) (March 1990):9–16.

Brown, R. A. "The Role of Scientists in Developing Hubble Space Telescope." *Optical Platforms, SPIE Proceedings,* 493:19. Edited by Charles L. Wyman. Bellingham, Wash.: Society for Photo-Optical Instrumentation Engineers, 1984.

Chaisson, E. C. "Early Results from the Hubble Space Telescope." *Scientific American*, 226(6) (1992):18–25.

Chaisson, E. C., and R. Villard. "The Science Mission of the Hubble Space Telescope." *Vistas in Astronomy,* 33(2) (1990):105–141.

Chaisson, Eric J. *The Hubble Wars: Astrophysics Meets Astropolitics in the Two-Billion-Dollar Struggle over the Hubble Space Telescope*. New York: HarperCollins Publishers, 1994.

DeVorkin, David H., and Robert W. Smith. *Hubble: Imaging Space and Time*. With contributions by Elizabeth A. Kessler. Washington, D.C.: National Geographic, 2008; revised 2011.

Dickinson, Terence. *Hubble's Universe: Greatest Discoveries and Latest Images*. Toronto: Firefly Books, 2012.

Field, George, and Donald Goldsmith. *The Space Telescope*. Chicago: Contemporary Books, 1989.

Fischer, Daniel, and Hilmar Duerbeck. *Hubble: A New Window to the Universe,* translated by Helmut Jenkner and Douglas Duncan. New York: Copernicus, 1996.

Goodwin, Simon. *Hubble's Universe: A Portrait of Our Cosmos*. New York: Viking Penguin, 1997.

Jenkins, Dennis R., and Jorge R. Frank. *Servicing the Hubble Space Telescope: Space Shuttle Atlantis—2009*. North Branch, Minn.: Specialty Press, 2009.

Kanipe, Jeff. *Chasing Hubble's Shadows: The Search for Galaxies at the Edge of Time.* New York: Hill and Wang, 2006.

Kerrod, Robin. *Hubble: The Mirror on the Universe.* Toronto: Firefly Books Ltd., 2003.

Kessler, Elizabeth A. *Picturing the Cosmos: Hubble Space Telescope Images and the Astronomical Sublime*. Minneapolis, Minn.: University of Minnesota Press, 2012.

[Lanzerotti, Louis, chair.] *Assessment of Options for Extending the Life of the Hubble Space Telescope: Final Report.* Committee on the Assessment of Options for Extending the Life of the Hubble Space Telescope, Space Studies Board, Aeronautics and Space Engineering Board, Division on Engineering and Physical Sciences, National Research Council of the National Academies. Washington, D.C.: National Academies Press, 2005.

Livio, Mario, Keith Noll, Massimo Stiavelli, and Michael Fall, eds. *A Decade of Hubble Space Telescope Science*. New York: Cambridge University Press; 2003.

Livio, Mario S., Michael Fall, and Piero Madau, eds. *The Hubble Deep Field: Proceedings of the Space Telescope Science Institute Symposium, held in Baltimore, Maryland, May 6–9, 1997*. Cambridge and New York: Cambridge University Press, 1998.

Longair, M. S. *Alice and the Space Telescope*. Baltimore: Johns Hopkins University Press, 1989.

Longair, M. S., J. W. Warner, L. Spitzer, Jr., and J. E. Gunn. *Scientific Research with the Space Telescope: International Astronomical Union Colloquium No. 54, held at the Institute for Advanced Study, Princeton, N.J.,*

August 8–11, 1979. Opening remarks by L. Spitzer, Jr., and summary by J. E. Gunn. Washington, D.C.: NASA Conf. Publ., NASA CP-2111, 1979.

Macchetto, F. Duccio, ed. *The Impact of HST on European Astronomy, Astrophysics and Space Science Proceedings*. Chichester, UK: Springer-Praxis, 2010.

[NASA.] *The Hubble Space Telescope Optical Systems Failure Report: Proceedings of the 12th Aerospace Sciences Meeting held 30 January–1 February, 1974*. Washington, D.C.: National Aeronautics and Space Administration, 1990.

Petersen, Carolyn Collins, and John C. Brandt. *Hubble Vision: Further Adventures with the Hubble Space Telescope*. New York: Cambridge University Press, 1998.

Preston, Richard. *First Light: The Search for the Edge of the Universe*. New York: Random House, 1996.

Rector, Travis A., Zoltan G. Levay, Lisa M. Frattare, Jayanne English, and Kirk Pu'uohau-Pummill. "Image-Processing Techniques for the Creation of Presentation-Quality Astronomical Images." *The Astronomical Journal*, 133(2):598–611.

Sembach, Kenneth R. *Hubble's Science Legacy: Future Optical/Ultraviolet Astronomy from Space. Proceedings of a Workshop Held at University of Chicago, Chicago, Illinois, USA, 2–5 April 2002*. San Francisco, Calif.: Astronomical Society of the Pacific, 2003.

Smith, Robert W. *The Space Telescope: A Study of NASA, Science, Technology, and Politics*. With contributions by Paul A. Hanle, Robert H. Kargon, and Joseph N. Tatarewicz. New York: Cambridge University Press, 1989.

Spitzer, Lyman, Jr. "Astronomical Research with the Large Space Telescope." *Science,* 161(3838) (1968):225–229.

———. "History of the Space Telescope." *Quarterly Journal of the Royal Astronomical Society,* 20(1979):29–36.

———. "Space Telescopes and Components." *Astronomical Journal,* 65(1960):242–263.

———. "The Space Telescope." *American Scientist,* 66 (July–August 1978):426–431.

U.S. Congress, House of Representatives, Committee on Science and Technology, Subcommittee on Space Science and Applications. *Space Telescope Program Review: Hearing before the Subcommittee on*

Space Science and Applications of the Committee on Science and Technology. 95th Congress, 2nd session, July 13, 1978. Washington, D.C.: U.S. Government Printing Office, 1978.

———. *Space Telescope, 1984: Hearing before the Subcommittee on Space Science and Applications of the Committee on Science and Technology*. 98th Congress, 2nd session, May 22, 24, 1984. Washington, D.C.: U.S. Government Printing Office, 1984.

U.S. Congress, House of Representatives, Committee on Science, Space, and Technology. *Space Telescope Flaw: Hearing before the Committee on Science, Space, and Technology*. 101st Congress, 2nd session, July 13, 1990. Washington, D.C.: U.S. Government Printing Office, 1990.

———. *Review of the Hubble Space Telescope Mission: Hearing before the Committee on Science, Space, and Technology*. 103rd Congress, 2nd session, March 8, 1994. Washington, D.C.: U.S. Government Printing Office, 1994.

U.S. Congress, Senate, Committee on Commerce, Science, and Transportation, Subcommittee on Science, Technology, and Space. *Hubble Space Telescope and the Space Shuttle Problems: Hearing before the Subcommittee on Science, Technology, and Space of the Committee on Commerce, Science, and Transportation*. 101st Congress, 2nd session, July 10, 1990. Washington, D.C.: U.S. Government Printing Office, 1990.

van den Bergh, S. "Astronomy with the Hubble Space Telescope." *Journal of the Royal Astronomical Society of Canada,* 79 (June 1985):134–142.

Weiler, Edward J. *Hubble: A Journey through Space and Time*. New York: Abrams, 2010.

Whitfield, Steven. *Hubble Space Telescope*. Burlington, Ontario: Apogee Books, 2006.

Zimmerman, Robert. *The Universe in a Mirror: The Saga of the Hubble Space Telescope and the Visionaries Who Built It.* Princeton, N.J.: Princeton University Press, 2008.

About the Contributors

David H. DeVorkin is a senior curator at the National Air and Space Museum (NASM), Smithsonian Institution. He is co-author of *Hubble: Imaging Space and Time* (National Geographic, 2008, 2011) and many other books.

Steven J. Dick served as the chief historian for the National Aeronautics and Space Administration (NASA) and director of the NASA History Office from 2003–2009 after working as an astronomer and agency historian for the U.S. Naval Observatory. Dick has also contributed substantially to the historical study of astronomy, having served as chairman of the Historical Astronomy Division of the American Astronomical Society and president of the History of Astronomy Commission of the International Astronomical Union. He has numerous publications in the fields of astrobiology and the history of life in the universe and has been appointed to the second chair of astrobiology at the Library of Congress (2013).

John Grunsfeld, currently the NASA associate administrator for Space Science, is a veteran of five spaceflights. He also served as the deputy director of the Space Telescope Science Institute. As an astronaut Grunsfeld performed several servicing and upgrade missions for the Hubble Space Telescope (HST) and has been on missions using the Astro-2 Observatory to observe the ultraviolet spectra of faint astronomical radiation.

Elizabeth A. Kessler was an assistant professor in the Art and Art History Department of Ursinus College and is now a visiting lecturer at Stanford University. Kessler's research focuses on the visual culture of science and its relationship to art with a special interest in HST imaging. She has held fellowships at Stanford University and NASM. Kessler's work includes her book *Picturing the Cosmos: Hubble Space Telescope Images and the Astronomical Sublime* as well as contributions to *Hubble: Imaging Space and Time*.

Roger D. Launius is a senior curator at NASM, Smithsonian Institution, and former NASA chief historian. He is author of *Space Stations: Base Camps to the Stars* (Smithsonian, 2009, rev. ed.) and many other books. He is now (2013) associate director for Collections and Curatorial Affairs at NASM.

David Leckrone came to the Goddard Space Flight Center in 1969 as an astrophysicist to work on the Orbiting Astronomical Observatory missions. Leckrone was then appointed scientific instruments scientist for the Space Telescope Program (before it became the HST Program) and head of the Astronomy Branch in the Laboratory for Astronomy and Solar Physics at Goddard. He has also been senior project scientist for the Goddard HST program from 1992 to 2009, where he helped lead five successful HST servicing missions. Leckrone's extensive publications include work in ultraviolet astronomy and the spectroscopic analysis of stellar atmospheres.

Zoltan Levay trained in photoelectric astronomy and has worked at the Space Telescope Science Institute (STScI) since 1983 to create software for astronomers to analyze the data obtained by the HST. Before joining STScI, Levay worked at the Computer Sciences Corporation and at Goddard Space Flight Center, where he supported various science missions, including the International Ultraviolet Explorer. Levay is currently working in the Office of Public Outreach at STScI as part of the Hubble Heritage Team; he transforms HST data into images and creates videos and other forms of media for public and educational consumption.

C. Robert O'Dell is distinguished research scientist in the Physics and Astronomy Department at Vanderbilt University. As an expert in the Orion and Helix nebulae, he joined NASA as a project scientist in 1972 after being director of the Yerkes Observatory of the University of Chicago. He remained with NASA throughout the "pre-HST" phase when the preliminary HST designs and project pitches to the Congress were being made.

Harold Reitsema is an astronomer and key member of the Ball Aerospace and Technologies Corp. staff developing many of the scientific instruments flown on the HST. In particular he was called upon to build the COSTAR (Corrective Optics Space Telescope Axial Replacement) device, the instrument flown on the first servicing mission to compensate for the flawed mirror for all instruments other than the Wide Field Planetary Camera (WFPC). After working for Ball he became a consultant to science institutions and NASA in the field of aerospace mission planning and development.

Nancy Grace Roman is a pioneer not only in astronomy but for women in science as well. After working in radio astronomy at the U.S. Naval Research Observatory, Roman joined NASA in early 1959 where she created the first NASA astronomical program and was named first chief of astronomy in the Office of Space Science. As director of this program, Roman was responsible for the launch of several X-ray, ultraviolet, and gamma-ray based satellite observatories.

Kenneth Sembach, head of the Hubble Mission Office at the Space Telescope Science Institute, cares for the scientific, operational, and managerial health of the HST. He is an astrophysicist who has been active in studies of the interstellar medium and galactic structure. He has collected and analyzed a wide range of ultraviolet data from space-based missions ranging from the International Ultraviolet Explorer to the Far Ultraviolet Spectroscopic Explorer, as well as the HST. He was the 2001 recipient of the Newton Lacy Pierce Prize in Astronomy, awarded to an astronomer under the age of 36 who has made highly important contributions to observational astronomy.

Robert W. Smith is a professor of History and Classics at the University of Alberta. His principal research interests are in the history of the physical sciences with a focus on astronomy, cosmology, and spaceflight. He is the author of the prize-winning history, *The Space Telescope: A Study of NASA, Science, Technology, and Politics* (Cambridge University Press, 1989); *The Expanding Universe: Astronomy's 'Great Debate' 1900–1931* (Cambridge University Press, 1982); and several other works.

Joseph N. Tatarewicz holds MA degrees in Philosophy (Catholic University 1976) and History and Philosophy of Science (Indiana University 1980) and a Ph.D. in History and Philosophy of Science (Indiana University 1984). He is the author of *Planetary Astronomy and Space Technology* (Indiana University Press,1990). He is presently an associate professor of history at the University of Maryland, Baltimore County.

John Trauger is a senior research scientist with the Jet Propulsion Laboratory and was principal investigator in the design and construction of the Wide Field and Planetary Camera 2, installed on the HST in 1993. Trauger was also a member of the NASA science working group for the WFC3, which was installed on the HST in 2009. His research is focused on atmospheres and magnetospheres of the outer planets in our solar system and the development of new methods for the direct imaging of planetary systems orbiting other stars in our galaxy.

Edward J. Weiler recently retired as NASA's associate administrator for Space Science. He also spent more than two decades as chief scientist for the HST. Prior to that post he served at the Goddard Space Flight Center as the director of space operations of the Orbiting Astronomical Observatory 3 (Copernicus) and went on to become the chief of the Ultraviolet–Visible and Gravitational Astrophysics Division at NASA Headquarters. He is author of the recent popular work *Hubble: A Journey through Space and Time* (Harry N. Abrams, 2010).

Index

Page numbers in *italics* refer to illustrations and captions.